MINITAB® MANUAL

DOROTHY WAKEFIELD
University of Connecticut Health Center

KATHLEEN MCLAUGHLIN
University of Connecticut

ELEMENTARY STATISTICS:
PICTURING THE WORLD
FIFTH EDITION

Ron Larson
Pennsylvania State University

Betsy Farber
Bucks County Community College

Prentice Hall
is an imprint of

Reproduced by Pearson Prentice Hall from electronic files supplied by the author.

Copyright © 2012, 2009, 2006 Pearson Education, Inc.
Publishing as Prentice Hall, 75 Arlington Street, Boston, MA 02116.

ISBN-13: 978-0-321-69377-8
ISBN-10: 0-321-69377-9

1 2 3 4 5 6 BRR 15 14 13 12 11

Prentice Hall
is an imprint of

www.pearsonhighered.com

▶ Introduction

The MINITAB Manual is one of a series of companion technology manuals that provide hands-on technology assistance to users of Larson/Farber *Elementary Statistics: Picturing the World,* Fifth Edition.

Detailed instructions for working selected examples, exercises, and Technology Labs from *Elementary Statistics: Picturing the World* are provided in this manual. To make the correlation with the text as seamless as possible, the table of contents includes page references for both the Larson/Farber text and this manual.

All of the data sets referenced in this manual are found on the data disk packaged in the back of every new copy of *Elementary Statistics: Picturing the World.* If needed, the MINITAB files (.mtp) may also be downloaded from the website www.pearsonhighered.com/mathstatsresources/.

iv

▶ **Contents**

Getting Started with MINITAB

▸ Using MINITAB Files

MINITAB is a Windows-based Statistical software package. It is very easy to use, and can perform many statistical analyses. When you first open MINITAB, the screen is divided into two parts. The top half is called the Session Window. The results of the statistical analyses are often displayed in the Session Window. The bottom half of the screen is the Data Window. It is called a Worksheet and will contain the data.

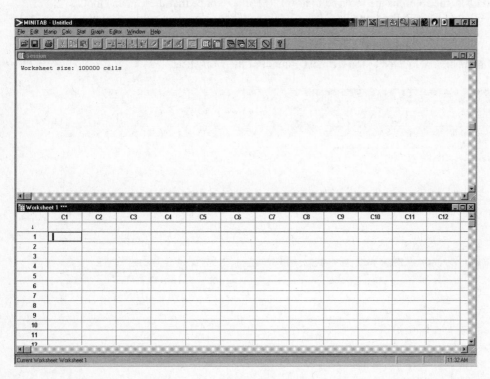

The data can either be entered directly into the Worksheet, or saved worksheets can be opened and used.

1

▸ Entering Data into the Data Window

To enter the data into the Data Window, you must first click on the bottom half of the screen to make the Data Window active. You can tell which half of the screen is active by the blue bar going across the screen. In the previous picture, notice that the blue bar is in the middle of the screen, highlighting **Worksheet 1.** This indicates that the Data Window is active. The bar will be gray if the Window is not active. (Notice the Session Window bar is gray.)

In MINITAB, the columns are referred to as C1, C2, etc. Notice that there is an empty cell directly below each heading C1, C2, etc. This cell is for a column name. Column names are optional because you can refer to a column as C1 or C2, but a name helps to describe the data contained in a column. Enter the data beginning in cell 1. Notice that the cell numbers are located in the leftmost column of the worksheet.

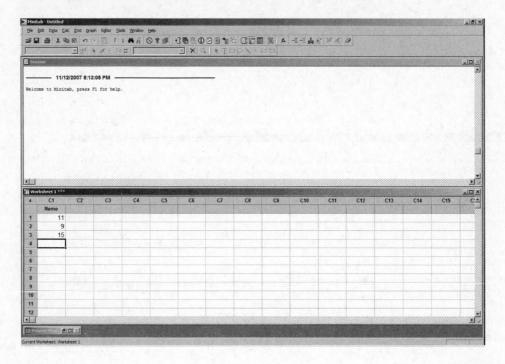

▶ Opening Saved Worksheets

Many of the worksheets that you will be using are saved on the enclosed data disk. To open a saved worksheet, click on **File → Open Worksheet.** The following screen will appear.

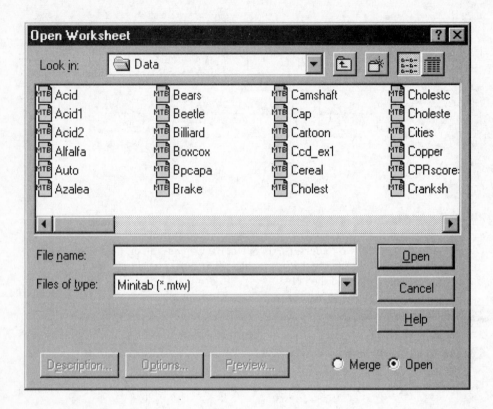

First, you must tell MINITAB where the data files are located. Since the data files are located on the data disk, you must tell MINITAB to **Look In** the **DVD** or **Compact Disc (D:).** To do this, click on the down arrow to the right of the top input field and select your DVD/CD drive by double-clicking on it.

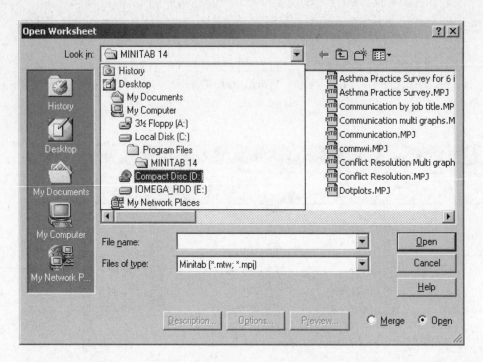

When you do this, you should see two folders listed. Select the Content folder, then select Data_Sets, and then Minitab data sets with a double-click. Now you should see a folder for each of the eleven chapters of the book.

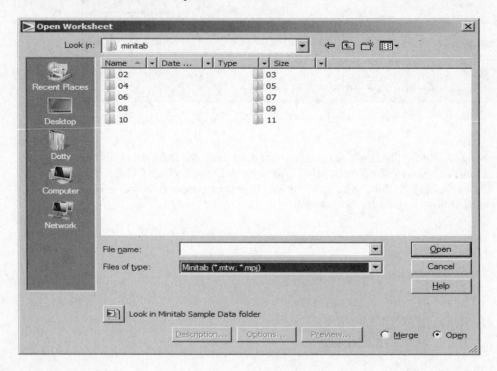

All data files are saved as MINITAB Portable worksheets and have the extension **.mtp.**
Click on the down arrow for the field called **Files of type** and select **Minitab Portable**
(*.mtp).

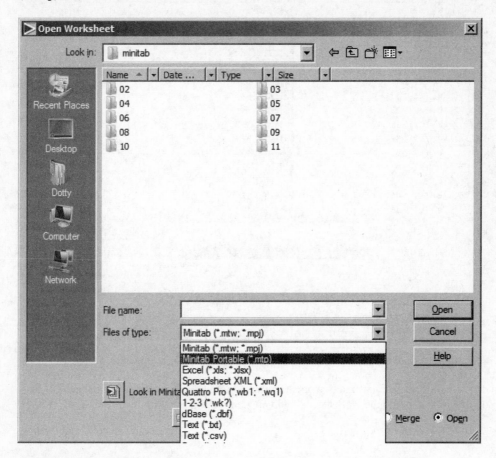

Now, select the folder called **02** (by double-clicking) and you should see all the
MINITAB worksheets for Chapter 2.

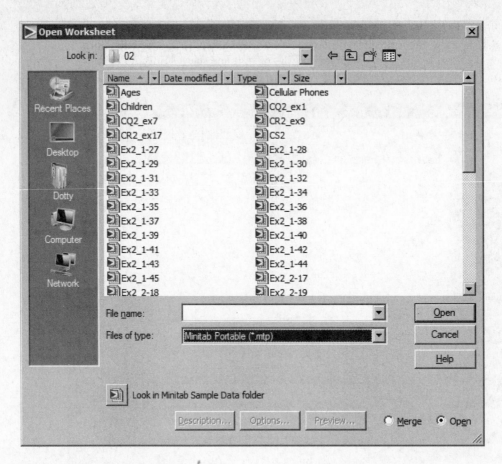

As you can see, the folder **02** has many worksheets saved to disk. To open the worksheet **Ages**, double-click on it and the worksheet should appear in the Data Window.

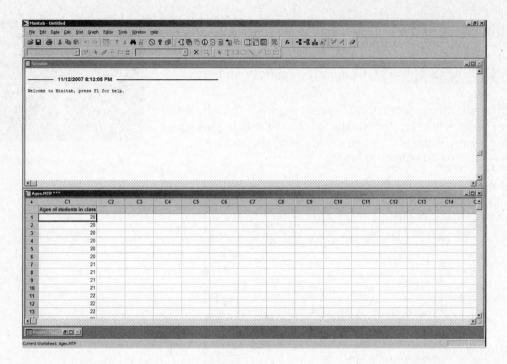

You are now ready to begin analyzing the data and learning more about MINITAB.

Introduction to Statistics

▶ Technology Exercises (pg. 35) | Generating random numbers

1. To select 10 randomly selected brokers from the company's 86, click on **Calc**
   → **Random Data** → **Integer.** The **Number of rows of data to generate** is
   10, and then **Store in column(s)** C1. The broker numbers will have a
   Minimum value 1 and **Maximum value** 86.

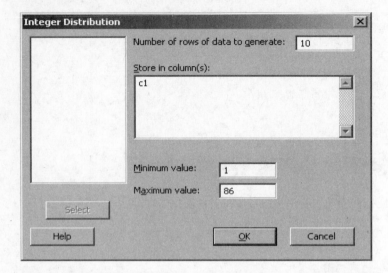

Click on **OK** and the 10 random numbers should be in C1 of the Data Window.

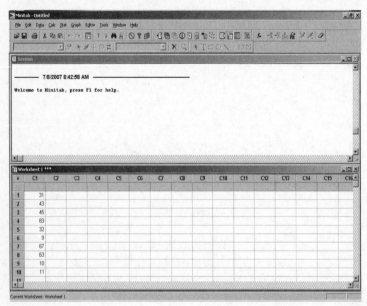

8

To order the sample list, click on **Data → Sort.** You should **Sort column:** C1, **By column:** C1 and **Store sorted data in → Original Column(s)**.

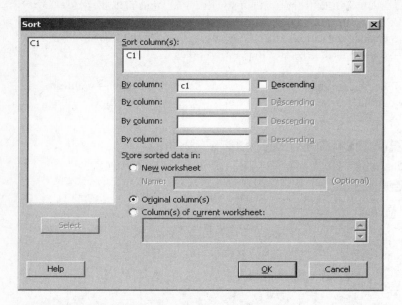

Click on **OK** and C1 should contain the sorted sample of 10 brokers.

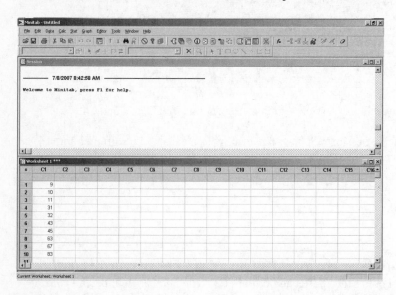

Since this is a *random* sample, each student will have different numbers in C1.

2. For this problem, use the same steps as above. Click on **Calc → Random Data → Integer.** The **Number of rows of data to generate** is 25, and then **Store in column(s)** C1. The camera phone numbers will have a **Minimum value** 1 and **Maximum value** 300. Finally, to order the sample list, click on **Data → Sort.** You should **Sort column:** C1, **By column:** C1 and **Store sorted data in → Original Column(s)**.

3. First store the integers 0 to 9 in C1 since you will want to be able to calculate an average later. Click on **Calc → Make Patterned Data → Simple Set of Numbers.** You should **Store patterned data in** C1. The numbers will begin **From the first value** 0 and go **To last value** 9 **In steps of** 1. Next, to randomly sample 5 digits, click on **Calc → Random Data → Sample from columns.** You need to **Sample** 5 **rows from column** C1 and **Store the sample in** C2. Repeat this two more times and **store the sample in** C3 and then in C4. Now you have generated the three samples.

    Now, to find the average of each of the four columns (C1, C2, C3, and C4), click on **Calc → Column Statistics.** The Statistic that you would like to calculate is the mean, so click on **Mean.** Enter C1 for the **Input variable** and click on **OK.**

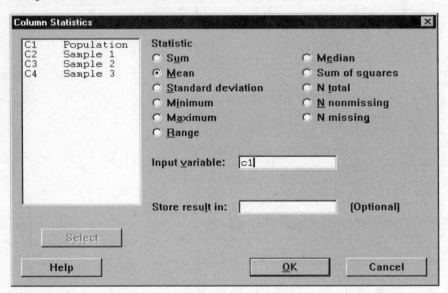

    The population mean will be displayed in the Session Window. Repeat this for C2, C3, and C4.

    The following output will appear in the Session Window; however, your means will be different for Columns 2 – 4 since these columns contain random data.

    Mean of Population = 4.5
    Mean of Sample 1 = 4.8
    Mean of Sample 2 = 3.8
    Mean of Sample 3 = 6.2

4. This problem will be very similar to the steps in problem 3. Click on **Calc → Make Patterned Data → Simple Set of Numbers.** You should **Store patterned data in** C1. The numbers will begin **From the first value** 0 and go **To last value** 40 **In steps of** 1. Next, to randomly sample 7 digits, click on **Calc → Random Data → Sample from columns.** You need to **Sample** 7 **rows from column** C1 and **Store the sample in** C2. Repeat this two more times and **store the sample in** C3 and then in C4. Now you have generated the three samples. Now, to find the average of each of the four columns (C1, C2, C3, and C4), click on **Calc → Column Statistics.** The Statistic that you would like to calculate is the mean, so click on **Mean.** Enter C1 for the **Input variable** and click on **OK.** The population mean will be displayed in the Session Window. Repeat this for C2, C3, and C4.

5. To simulate rolling a 6-sided die, you want to sample with replacement from the integers 1 to 6. You could enter the numbers 1 to 6 into C1 and then sample with replacement. A better way to do this is to click on **Calc → Random Data → Integer.** The **Number of rows of data to generate** is 60 and **Store in column** C1. This represents the 60 rolls. Enter a **Minimum value** of 1 and a **Maximum value** of 6 to represent the six sides of the die.

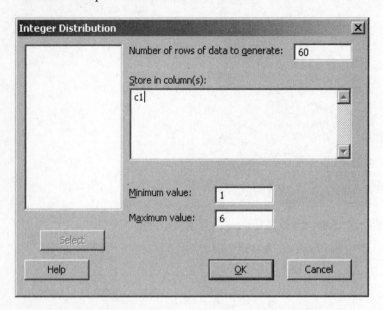

Click on **OK** and C1 should have the results of 60 rolls of a die.

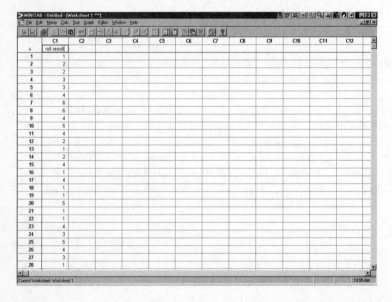

To count how many times each number was rolled, click on **Stat → Tables → Tally Individual Variables.** Enter C1 for the **Variable** and select **Counts** by clicking on it.

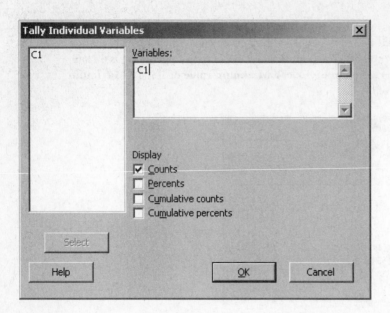

Click on **OK** and the totals will be displayed in the Session Window, as shown below.
Tally for Discrete Variables: C1
C1 Count
  1   10
  2   12
  3   8
  4   13
  5   9
  6   8
N=   60

Recall that each student's results will look different because this is random data. In the results above, 10 ones, 12 twos, 8 threes, 13 fours, 9 fives, and 8 sixes were rolled.

7. To simulate tossing a coin, you can repeat the steps in problem 5. Click on **Calc → Random Data → Integer.** You want to **Generate** 100 **rows of data** and **Store in column** C1. This represents the 100 tosses. Enter a **minimum value** of 0 and a **maximum value** of 1 to represent the two sides of the coin. Click on **OK** and C1 should have the results of 100 tosses of a coin. To count how many times each side of the coined was tossed, click on **Stat → Tables → Tally Individual Variables.** Enter C1 for the **Variable** and select **Counts** by clicking on it. Click on **OK** and the counts will be displayed in the Session Window. Recall that 0 represents heads and 1 represents tails.

Tally for Discrete Variables: C1
C1 Count
  0   49
  1   51
N=   100

In this example, 49 heads and 51 tails were rolled.   ◀

Descriptive Statistics

Section 2.1

▸ Example 7 (pg. 46) Construct a histogram using the GPS data

To create this histogram, you must enter the data table, found on page 41 of the textbook, into a Minitab worksheet. You will only need to enter the midpoints and frequencies. Label each column appropriately as shown below.

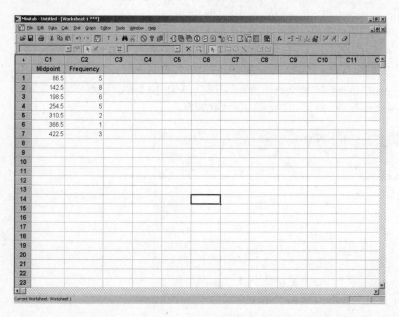

Now you are ready to make the histogram. Click on: **Graph → Histogram**. Select a **Simple** histogram by highlighting it and click OK.

13

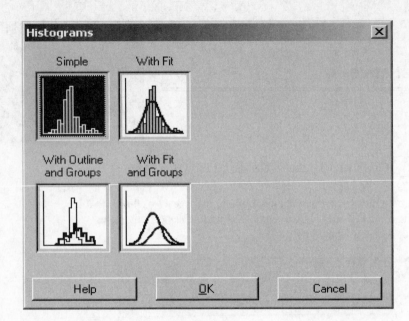

On the main Histogram screen, double-click on C1 in the large box at the left of the screen. "Midpoint" should now be filled in as the **Graph variable**.

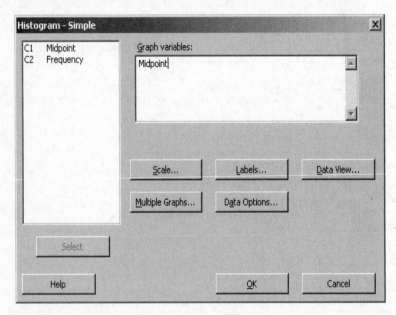

Next, click on **Data Options**, and then on the **Frequency** tab of the pop-up. Select C2 (Frequency) for the **Frequency variable**. This tells Minitab which column the frequencies are in when graphing the midpoints of the costs.

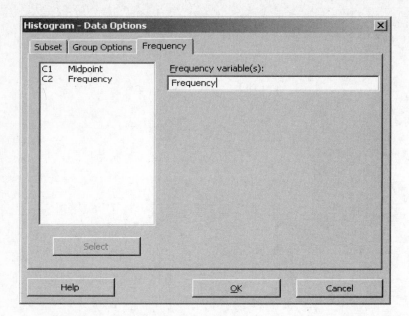

Click on **OK** to return to the main Histogram screen, and next click on the **Labels** button to enter a title for the graph. Click on **OK**.

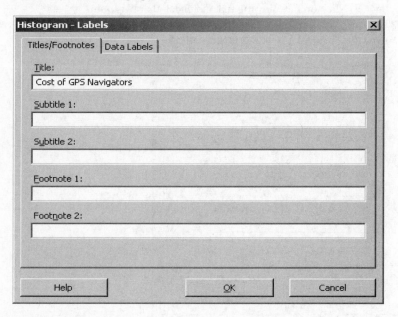

At this point, if you click on **OK** again, MINITAB will draw a histogram using default settings. Your histogram will look like the one below.

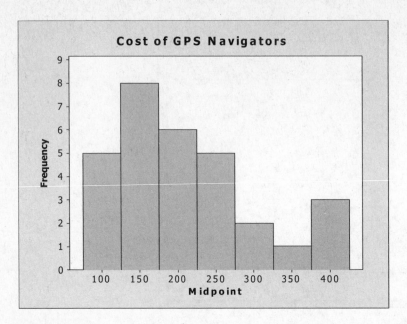

Notice that the numbering along the axis is not the midpoints. We can edit the graph to fix this, however. On the graph window, right-click on the X-axis and select **Edit X Scale** from the drop-down menu. Click on the **Binning** tab and check that **Midpoint** is selected. Enter "86.5 : 422.5 / 56" for **Midpoint/Cutpoint Positions**. This tells Minitab you want the numbering to go from from 86.5 to 422.5 in steps of 56.

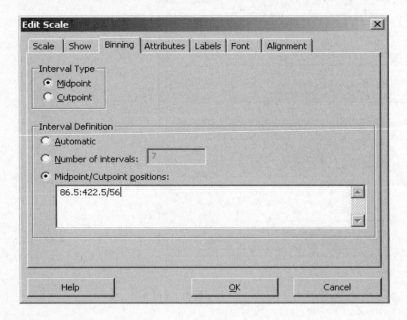

Click on **OK** and the changes should have been made to graph. Next change the X-axis label to "Cost of GPS Navigators". Right-click on the current axis label, and select "Edit X axis label" from the drop-down menu. Enter "Cost of GPS Navigators" below **Text**.

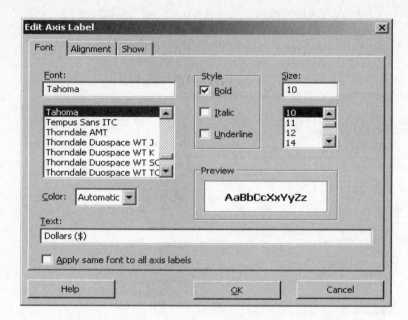

Click on **OK** to view the changes to your histogram.

Note: If you would like to add the frequencies above each rectangle of the histogram, right-click on a rectangle of the graph and select **Add → Data Labels.** Click on **Use y-value labels** and click on **OK** to view the changes.

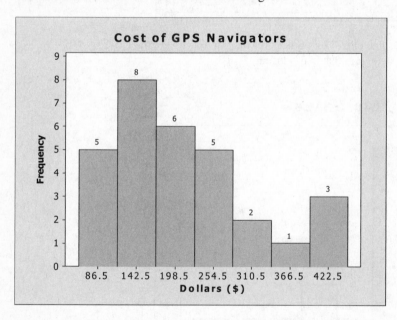

To print the graph, click on **File → Print Graph.** Next, click **OK** and the graph should print.

◄

▸ Exercise 31 (pg. 50)   Construct a frequency histogram using 6 Classes

Open the worksheet **ex2_1-31** which is found in the **ch02** MINITAB folder. Click on:
Graph → Histogram. Select a **Simple** histogram and click OK.
On the main Histogram screen, double-click on C1 in the large box at the left of the
screen. "July Sales" should now be filled in as the **Graph variable**.

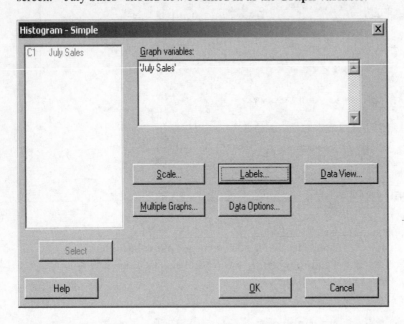

Click on the **Labels** button and enter "July Sales" for the title for the graph. Click on **OK**
twice to view the default histogram that is produced.

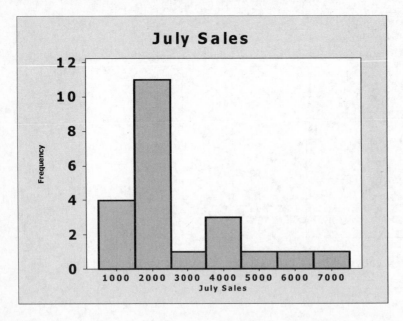

Now change the axis label to "Dollars". To do this, right-click on the axis label (July Sales), and select "Edit X axis label" from the drop-down menu. Type "Dollars" below **Text**. Click on **OK** to view the changes.

Next, decide on the numbering for the X-axis that is needed for the 6 classes. This will depend on the class limits you used in your frequency distribution. In order to use your class limits, you must tell MINITAB to use **Cutpoints.** To do this, right-click on the current X-axis numbering. Select "Edit X Scale" from the drop-down menu. Click on the **Binning** tab. Select **CutPoint** as the **Interval Type.** Next tell MINITAB what the cutpoint positions are. One solution is to use cutpoints beginning at 1000 and going up to 7600 in steps of 1100. So fill in 1000:7600/1100 for **Midpoint/cutpoint positions**.

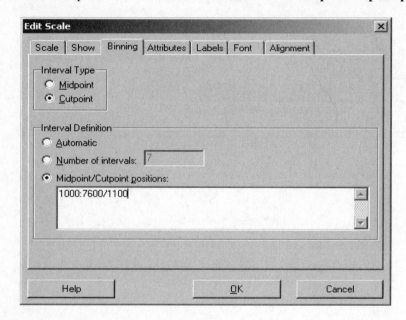

Click on **OK** to view the changes.

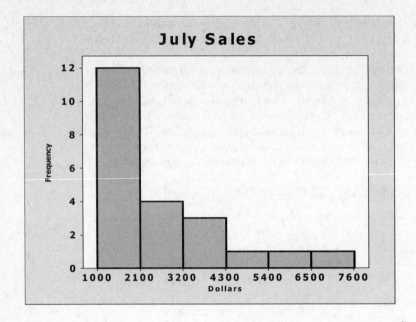

▶ Exercise 35 (pg.50)   Gasoline consumption (mpg)

Open worksheet **EX2_1-35** which is found in the **ch02** MINITAB folder. Click on:
Graph → Histogram. Select a **Simple** histogram and click OK.
On the main Histogram screen, double-click on C1 in the large box at the left of the
screen. "Fuel Consumptions (in mpg)" should now be filled in as the **Graph variable**.
Click on the **Labels** button and enter "Fuel Consumption" for the title for the graph.
Click on **OK** twice to view the default histogram that is produced.

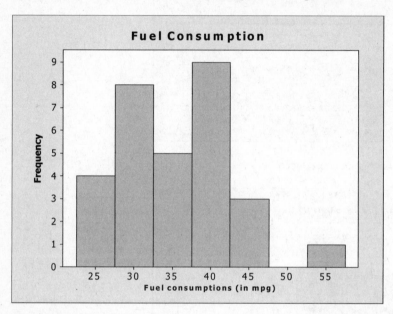

Next, change the label for the X-axis. To do this, right-click on the axis label, and select
"Edit X axis label" from the drop-down menu. Type "Miles per Gallon" below **Text**.
Click on **OK** to view the changes.

Now, decide on the numbering for the X-axis that is needed for the 5 classes. You can
either choose the classes yourself as in the last example, or let MINITAB do it for you.
Right-click on the X-axis numbers and select "Edit X Scale" from the drop-down menu.
Click on the **Binning** tab. Select **CutPoint** as the **Interval Type.** Next tell MINITAB
that the **Midpoint/Cutpoint positions** are 24:59/7. Click on OK.

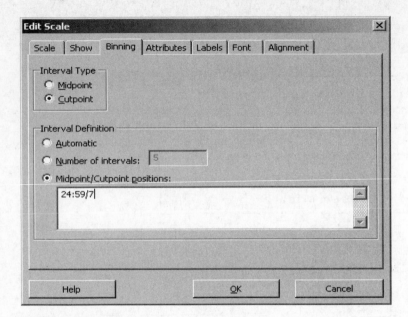

The following frequency histogram is produced.

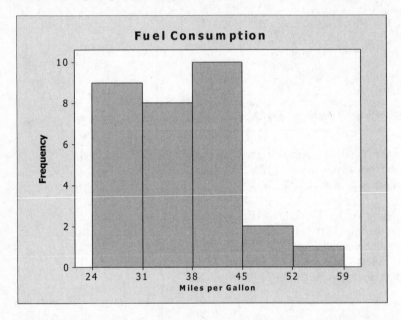

Finally, since this is a relative frequency histogram, the right-click on the Y-axis numbering, and select "Edit Y Scale" from the drop-down menu. Click on the **Type** tab and select **Percent.** Click on **OK** twice, and the histogram should have all of the changes.

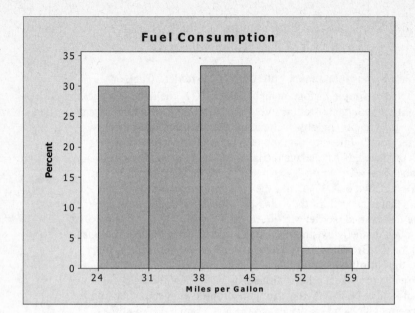

▸ Exercise 36 (pg. 50) Construct a relative frequency histogram
for the ATM data using 5 classes

Open worksheet **ex2_1-36** which is found in the **ch02** MINITAB folder. Click on:
Graph → Histogram. Select a **Simple** histogram and click OK. On the main Histogram
screen, double-click on C1 in the large box at the left of the screen. "ATM withdrawals"
should now be filled in as the **Graph variable**. Click on the **Labels** button and enter
"Daily ATM Withdrawals" for the title for the graph. Click on **OK** twice to view the
default histogram that is produced. Next, change the label for the X-axis. To do this,
right-click on the axis label, and select "Edit X axis label" from the drop-down menu.
Type "Dollars" below **Text**. Click on **OK** to view the changes. Now, decide on the
numbering for the X-axis that is needed for the 5 classes. You can either choose the
classes yourself as in the last example, or let MINITAB do it for you. In this example, let
MINITAB do it. Right-click on the X-axis numbers and select "Edit X Scale" from the
drop-down menu. Click on the **Binning** tab. Select **CutPoint** as the **Interval Type**.
Next tell MINITAB that the **Number of Intervals** is 5. Click on **OK** to view the
frequency histogram. Finally, to make a relative frequency histogram, the right-click on
the Y-axis numbering, and select "Edit Y Scale" from the drop-down menu. Click on the
Type tab and select **Percent**. Click on **OK** twice, and the histogram have all of the
changes.

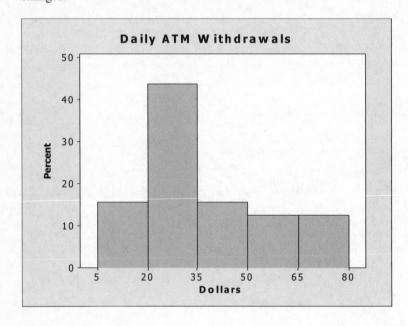

Section 2.2

▶ Example 2 (pg. 54) Constructing a stem-and-leaf plot

Open the file **Text Messages** which is found in the **ch02** MINITAB folder. This
worksheet contains the numbers of text messages sent last month by cellular phone users.
The data should appear in C1 of your worksheet.

To construct a Stem-and-leaf plot, click on **Graph → Stem-and-Leaf.**
On the screen that appears, select C1 as your **Variable** by doubling clicking on C1. Click
on **OK.**

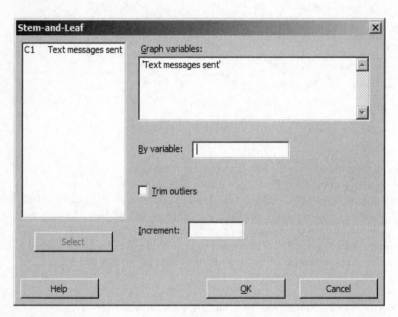

The stem-and-leaf plot will be displayed in the Session Window.

Stem-and-Leaf Display: Messages
Stem-and-leaf of Messages N = 50
Leaf Unit = 1.0

```
 1   7   8
 1   8
 1   8
 1   9
 1   9
 1  10
 6  10   58999
11  11   22234
19  11   67888999
(8) 12   11222344
23  12   6666699
16  13   02334
11  13   99
 9  14   024
 6  14   5578
 2  15
 2  15   59
```

In this MINITAB display, the first column on the left is a counter. This column counts the number of data points starting from the smallest value (at the top of the plot) down to the median. It also counts from the largest data value (at the bottom of the plot) up to the median. Notice that there is only one data point in the first row of the stem-and-leaf. There are no data points in rows 2 or 6, so the counter on the left remains at "1". Row 7 has 5 data points so the counter increases to "6". The row that contains the median has the number "8" in parentheses. This number counts the number of data points that are in the row that contains the median.

The second column in the display is the **Stem**. In this example, the Stem values range from 7 to 15. Notice that this display contains two rows for each of the values from 8 through 15. These are called **split-stems**. For each stem value, the first row contains all data points with leaf values from 0 to 4 and the second row contains all data points with leaf values from 5 to 9. Notice that MINITAB constructs an *ordered* stem-and-leaf.

The leaf values are shown to the right of the stem. The leaf values may be the actual data points or they may be the rounded data points. To find the actual values of the data points in the display, use the "Leaf Unit=" statement at the top of the display. The "Leaf Unit" gives you the place value of the leaves. In this stem-and-leaf, the first data point has a stem value of 7 and a leaf value of 8. Since the "Leaf Unit=1.0", the 8 is the "ones" place and the 7 is in the "tens" places, thus the data point is 78.

◄

▶ Example 3 (pg. 55) Constructing a dotplot

Open the file **Text Messages** which is found in the **ch02** MINITAB folder. To construct a dotplot, click on **Graph → Dotplot → Simple.** Select C1 for **Graph Variables.** Click on the **Labels** button, enter an appropriate **Title** and click on **OK.**

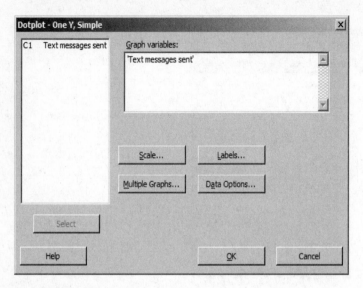

The following dotplot should appear.

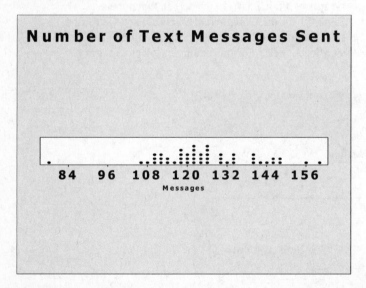

Each dot in the above plot represents the text messages for an individual caller. For example, one caller sent 78 messages, one sent 108, and 3 sent 109 messages (the 3 dots above 109). You can edit the numbering along the X-axis just as in the Histograms.

▶ Example 4 (pg. 56) Constructing a pie chart

In this example, you must enter the data into the Data Window. Begin with a clean
worksheet. From the table in the left margin of page 56 of the textbook, enter the Degree
types into C1 and the number of degrees into C2. Label each column appropriately as
shown below.

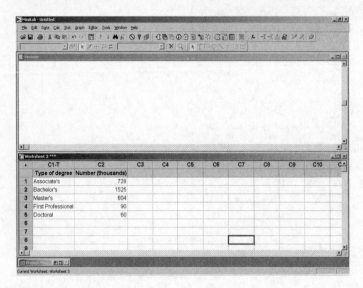

To construct the pie chart, click on **Graph → Pie Chart**. In the screen that appears,
select **Chart values from a table**. Enter C1 for the **Categorical Variable** and C2 for the
Summary Variable. Click on the **Labels** button and enter an appropriate title. Once in
the **Labels** screen, you can also select **Slice Labels.** For this example, select **Category
Name** and **Percent**.

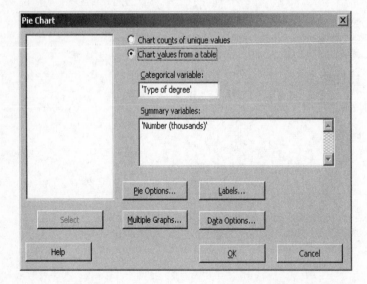

Click on **OK** to view the pie chart.

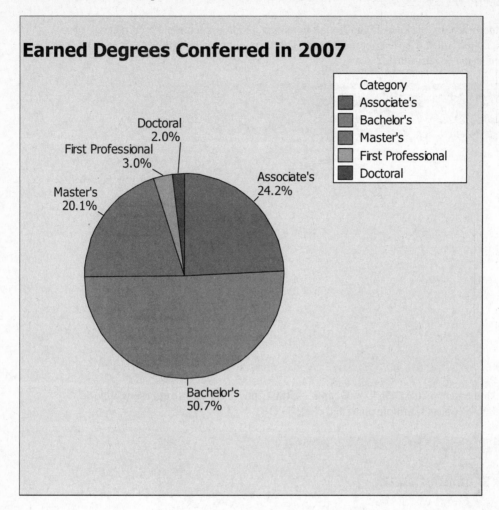

▶ Example 5 (pg. 57) Constructing Pareto charts

Enter the Inventory Shrinkage data (found in the paragraph for Example 5 on page 57 of the text) into C1 and C2. Do not include the Total amount of $36.5 million. Note: Do NOT enter the $ signs into C2.

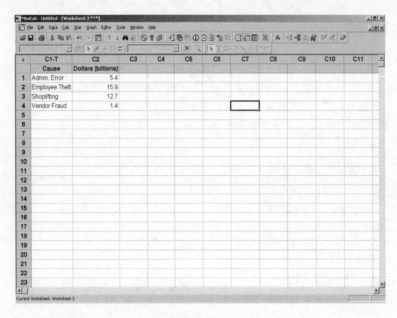

To make the Pareto chart, click on **Graph → Bar Chart**. The **Bars represent** Values from a table. Select a **Simple** chart and click on **OK**.

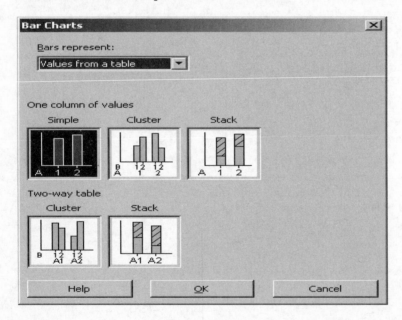

Enter C2 for **Graph variables** and C1 for **Categorical variable.** Click on **Bar chart options** and select **Decreasing Y.** Click on the **Labels** button and enter an appropriate title. Now view the chart.

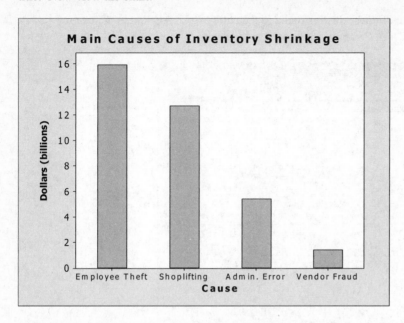

Note: Notice that the above chart has spaces between the rectangles. To remove these spaces, just edit the graph. Right-click on the X-axis and select "Edit X Scale" from the drop-down menu. On the **Scale** tab, click on **Gap between clusters** and enter a 0. This will produce a chart with no space between bars.

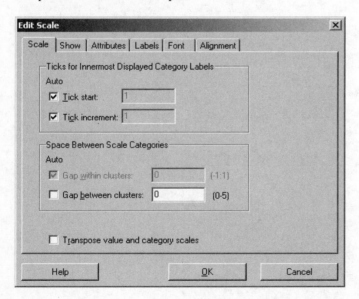

▶ Example 7 (pg. 59) Construct a time series chart of cellular
telephone subscribers

Open worksheet **Cellular Phones** which is found in the **ch02** MINITAB folder. Click on
Graph → Time Series Plot → Simple. Select C2 as the **Series**.

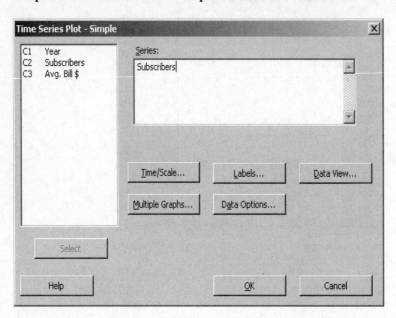

Click on the **Time/Scale** button. Select **Stamp** and enter **C1** (Year) for the **Stamp
Columns.** Click on **OK**.

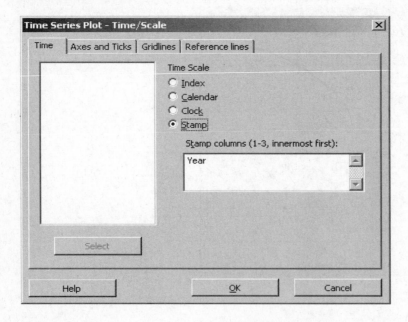

Click on the **Labels** button and enter an appropriate title for the plot.

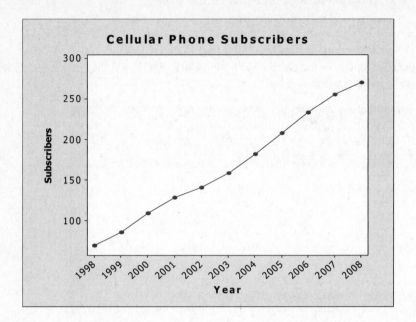

▶ Exercise 17 (pg. 61) Construct a stem-and-leaf of the exam
scores on a biology midterm.

Open worksheet **ex2_2-17** which is found in the **ch02** MINITAB folder. Click on
Graph→ Stem-and-Leaf. Select C1 for the **Graph Variable.** Click on **OK** and the
stem-and-leaf plot should be in the Session Window.

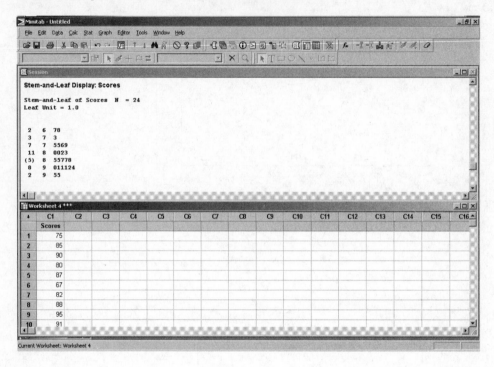

▶ Exercise 22 (pg. 62)   Construct a dotplot of the lifespan (in days)
of houseflies

Open worksheet **ex2_2-22** which is found in the **ch02** MINITAB folder. Click on **Graph**
→ **Dotplot** → **Simple.** Select C1 for the **Graph Variable.** Click on the **Labels** button
and enter an appropriate **Title.** Click on **OK** to view the dotplot.

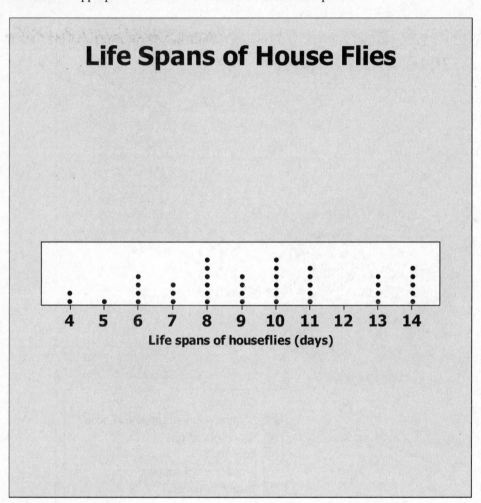

Life Spans of House Flies

Life spans of houseflies (days)

▶ Exercise 24 (pg. 62) Construct a pie chart of the data.

This data must be entered into the Data Window. Enter the categories into C1 and the Expenditures into C2. Click on **Graph → Pie Chart.** Select **Chart values from a table.** Enter C1 for the **Categorical Variable** and C2 for the **Summary Variable.** Click on the **Labels** button, enter an appropriate **Title** and click on **OK.** (You can also select **Slice Labels.**)

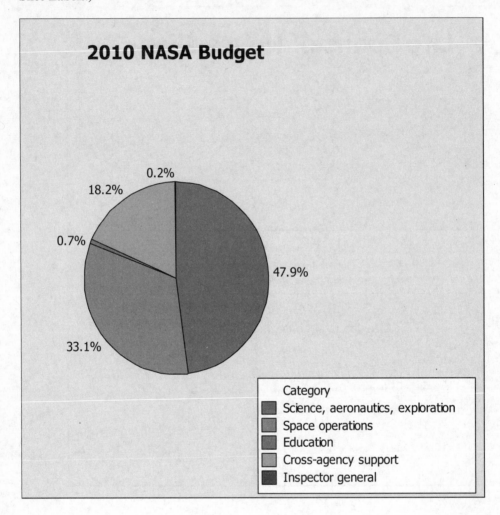

2010 NASA Budget

Category
Science, aeronautics, exploration
Space operations
Education
Cross-agency support
Inspector general

▶ Exercise 26 (pg. 62) Construct a Pareto chart to display the data

Enter the data into the Data Window. Enter the cities into C1 and the ultraviolet index
for each city into C2. To make the Pareto chart, click on **Graph → Bar Chart**. The
Bars represent Values from a table. Select a **Simple** chart and click on **OK**. Enter C2
for **Graph variables** and C1 for **Categorical variable.** Click on **Bar chart options** and
select **Decreasing Y.** Click on the **Labels** button and enter an appropriate title. Now
view the default chart. Notice that the default chart has spaces between the rectangles.
To remove these spaces, just edit the graph. Right-click on the X-axis and select "Edit X
Scale" from the drop-down menu. On the **Scale** tab, click on **Gap between clusters** and
enter a 0. Click on **OK** to view the Pareto chart.

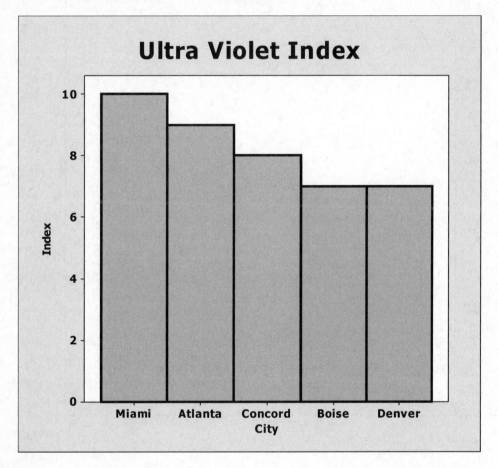

▶ Exercise 28 (pg. 63) Construct a scatterplot of the data.

Enter the Number of Students per Teacher into C1 and the Average Teacher's Salary into C2 (data found to the left of exercise). Click on **Graph → Scatterplot → Simple.** Select C2 for the **Y variable** and C1 for the **X variable.** Click on the **Labels** button and enter an appropriate title. Click on **OK** twice to view the scatterplot.

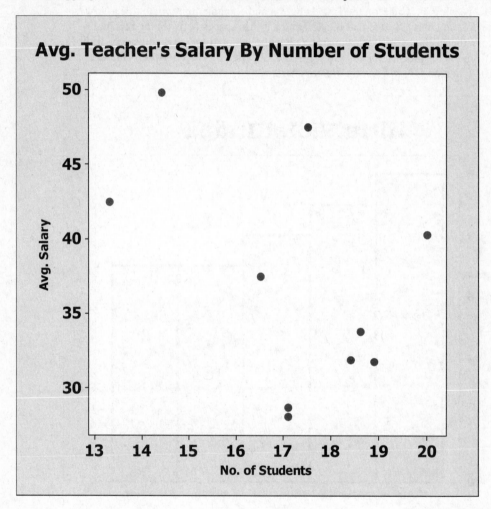

▸ Exercise 29 (pg. 63) Construct a time series plot of temperatures.

Enter the data into the Data Window. Enter the date into C1 and the temperature into C2.
Click on **Graph → Time Series Plot → Simple.** Select C2 as the **Series.** Click on the
Time/Scale button. Select **Stamp** and enter C1 (Year) for the **Stamp Columns.** Click
on **OK.** Click on the **Labels** button and enter an appropriate title for the plot. Click on
OK to view the plot.

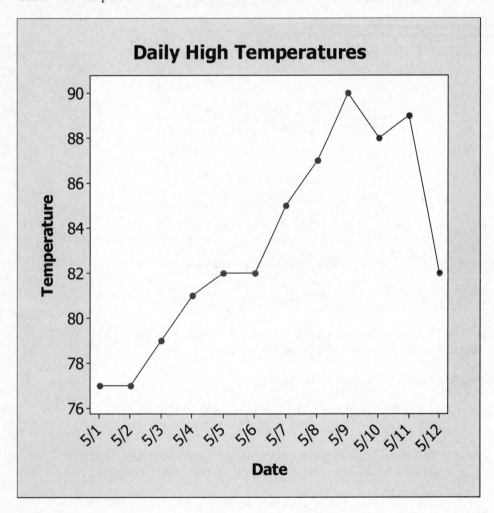

Section 2.3

▶ Example 6 (pg. 68) Find the mean and standard deviation of the
age of students

Finding the mean and standard deviation of a dataset is very easy using MINITAB. Open the worksheet **Ages** which is found in the **ch02** MINITAB folder. Click on **Stat → Basic Statistics → Display Descriptive Statistics.** You should see the input screen below.

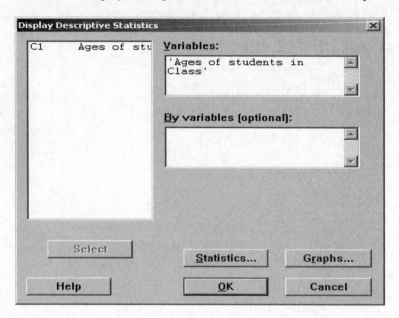

Double click on C1 to select the age data that is entered in C1. Click on **OK** and the descriptive statistics should appear in the Session Window.

Descriptive Statistics: Ages of students in class

Variable	N	Mean	StDev	Minimum	Q1	Median	Q3	Maximum
Ages	20	23.75	9.81	20.00	20.00	21.50	23.00	65.00

Notice that MINITAB displays several descriptive statistics: sample size, mean, standard deviation, minimum value, maximum value, median, and the first and third quartiles.

The mode is NOT produced by the above procedure; however, it is quite simple to have MINITAB tally up the data values for you, and then you can select the one with the highest count. Click on **Stat → Tables → Tally Individual Variables**. On the input screen, double-click on C1 to select it. Also, click on **Counts** to have MINITAB count up the frequencies for you.

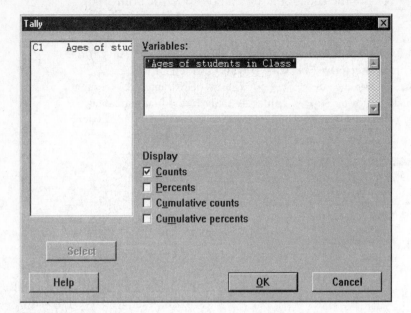

When you click on **OK**, a frequency table will appear in the Session Window.

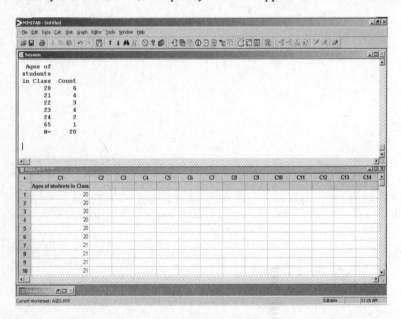

Notice that Age 20 has a count of 6. This means that 6 people in the data were age 20.
Since this is the highest count, 20 is the mode.

To print the Session Window with both the descriptive statistics and the frequency table
in it, click anywhere up in the Session Window to be sure that it is the active window.
Next click on **File → Print Session Window**.

▸ Exercise 21 (pg. 73) Find the mean, median, and mode for points per
game scored by each NFL team

Open worksheet **ex2_3-21** which is found in the **ch02** MINITAB folder. Click on **Stat→**
Basic Statistics → Display Descriptive Statistics. Double-click on C1 to select it.
Click on **OK** and the results should be in the Session Window. Next, make the frequency
table to help find the mode. Click on **Stat → Tables → Tally Individual Variables**.
Double-click on C1, then click on **OK**. Now both of the displays will be in the Session
Window.

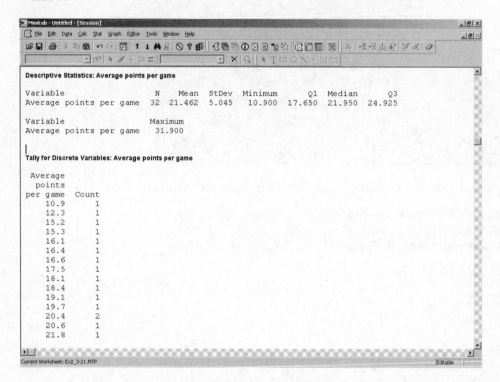

The mean number of points scored per team is 21.46 and the median is 21.95. The mode
is the points per team with the highest count. In this problem, the mode is 20.4. Notice
that it has a count of 2.

▶ Exercise 55 (pg. 77) Construct a frequency histogram of the
                        heights using 5 classes

Open worksheet **ex2_3-55** which is found in the **ch02** MINITAB folder. Click on **Graph**
→ **Histogram** → **Simple.** Select C1 for the **Graph variable.** Click on the **Labels**
button and enter an appropriate title. Click on OK to view the default histogram. Since
you need 5 classes, edit the graph. Right-click on the X-axis and select "Edit X scale"
from the drop-down menu. On the **Binning** tab, select **Cutpoints,** and enter 5 for the
Number of Intervals. Click on **OK** to view the changes to the graph.

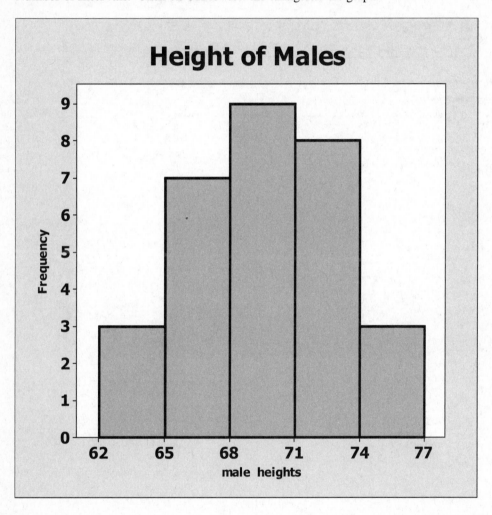

▶ Exercise 64 (pg. 78) Find the mean and median, and construct a
stem-and-leaf plot of the data

Open worksheet ex**2_3-64** which is found in the **ch02** MINITAB folder. Click on **Stat**
→ **Basic Statistics** → **Display Descriptive Statistics.** Select C1 for the **Variable** and
click on **OK.** The descriptive statistics should be in the Session Window. Next, click on
Graph → **Stem-and-Leaf.** Select C1 for the **Variable** and click on **OK**. The stem-and-
leaf plot should also be in the Session Window as shown in the next picture.

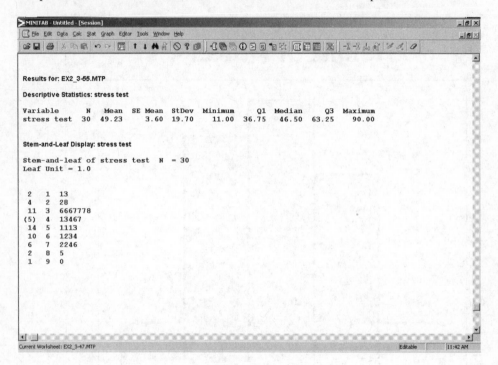

Section 2.4

▶ Example 5 (pg. 84) Calculate the mean and standard deviation

Open worksheet **rentrate** which is found in the **ch02** MINITAB folder. Click on **Stat →
Basic Statistics → Display Descriptive Statistics.** Select C1 for the **Variable** and click
on **OK.** The descriptive statistics should be in the Session Window.

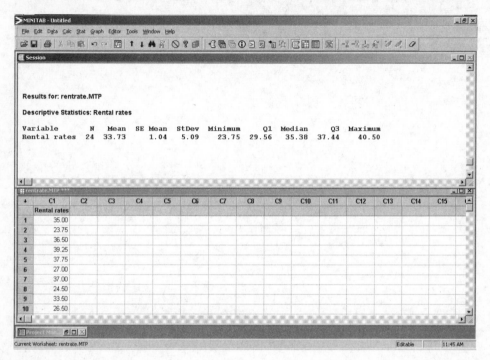

▶ Exercise 11 (pg. 90) Find the range, mean, variance, and standard
deviation of the dataset.

Enter the data into C1 of the Data Window. Click on **Stat → Basic Statistics → Display
Descriptive Statistics.** Select C1 for the **Variable** and click on **OK.** The descriptive
statistics should be in the Session Window.

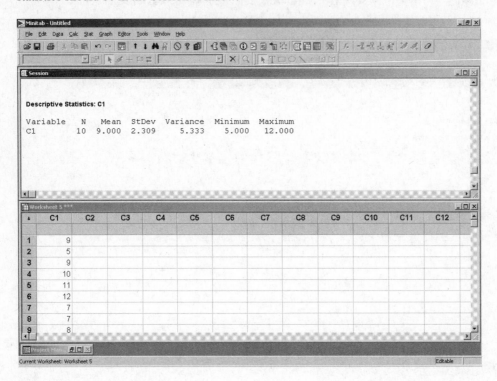

The range can be calculated by subtracting (Maximum – Minimum = 7) from the output.

◀

▶ Exercise 25 (pg. 92)   Find the mean, range, standard deviation,
and variance for each city

Enter the data into C1 and C2 of the Data Window. Be sure to label the columns. Click
on **Stat → Basic Statistics → Display Descriptive Statistics.** Select both C1 and C2 for
the **Variable** and click on **OK.** The descriptive statistics should be in the Session
Window.

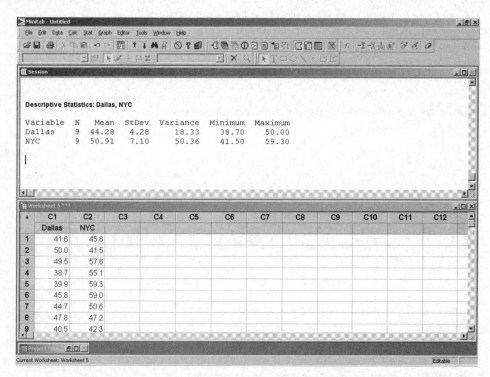

The average salary is for an accountant is higher in NYC ($50,910) compared to Dallas
($44, 280). Notice that the standard deviation of salaries in NYC is also higher,
indicating that there is more variability in salaries in NYC.

Section 2.5

▶ Example 2 (pg. 101)   Find the first, second, and third quartiles of
the tuition data

Open worksheet **Tuition** which is found in the **ch02** MINITAB folder. Click on **Stat →
Basic Statistics → Display Descriptive Statistics.** Select C1 for the **Variable** and click
on **OK.** The descriptive statistics should be in the Session Window. Recall that the
median is the second quartile.

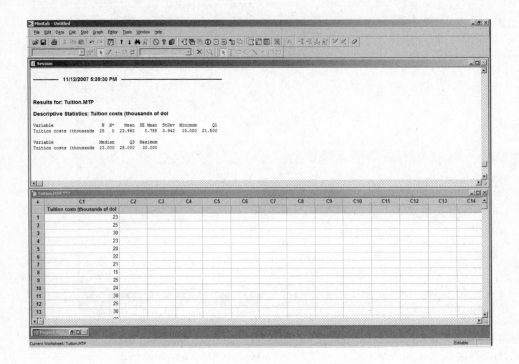

▶ Example 4 (pg. 103) Construct a box-and-whisker plot using the
                      data given in Example 1

Enter the number of nuclear power plants (data found on page 100 of the text) into C1 of
the Data Window. Click on **Graph → Boxplot → Simple**. Select C1 for the **Graph
variable.** Next, since by default MINITAB plots vertically, click on the **Scale** button and
select **Transpose value and category scales.** This will turn the plot horizontal, as in the
textbook. Click on the **Labels** button and enter an appropriate title. Click on **OK** twice to
view the boxplot.

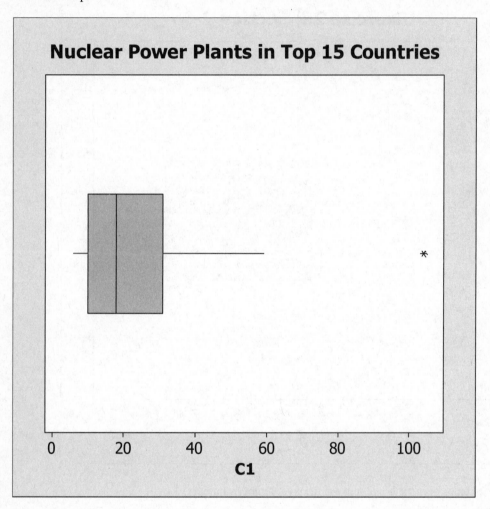

▶ Exercise 23 (pg. 108) Draw a boxplot of the data

Open worksheet **ex2_5-23** which is found in the **ch02** MINITAB folder. Click on **Graph** → **Boxplot** → **Simple**. Select C1 for the **Graph variable.** Next, since by default MINITAB plots vertically, click on the **Scale** button and select **Transpose value and category scales.** This will turn the plot horizontal, as in the textbook. Click on the **Labels** button and enter an appropriate title. Click on **OK** twice to view the boxplot.

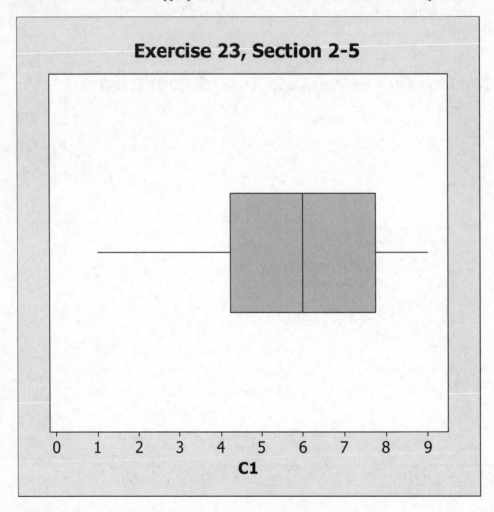

Hold the cursor over the boxplot to see the five-number summary.

▶ Exercise 31 (pg. 108) Draw a boxplot of hours of TV
watched per day

Open worksheet **ex2_5-31** which is found in the **ch02** MINITAB folder. Click on **Graph**
→ **Boxplot** → **Simple**. Select C1 for the **Graph variable**. Next, since by default
MINITAB plots vertically, click on the **Scale** button and select **Transpose value and
category scales.** This will turn the plot horizontal, as in the textbook. Click on the
Labels button and enter an appropriate title. Click on **OK** twice to view the boxplot.

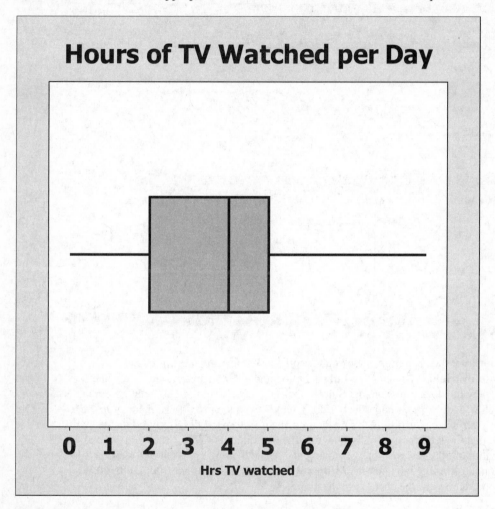

Hold the cursor over the boxplot to see the five-number summary.

▶ Technology Lab (pg. 121) Use descriptive statistics and a
histogram to describe the milk data

Open worksheet **Tech2** which is found in the **ch02** MINITAB folder. Click on **Stat** →
Basic Statistics → **Display Descriptive Statistics.** Click on **OK** and the descriptive
statistics will be displayed in your Session Window.

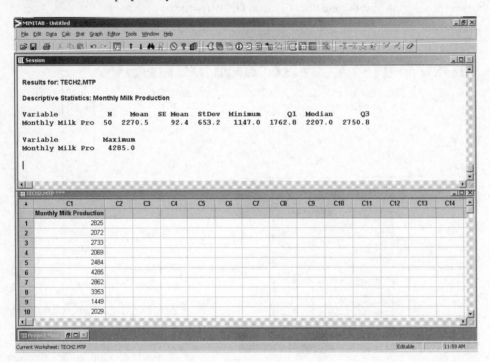

Next construct the histogram. Click on **Graph** → **Histogram** → **Simple.** Select C1 for
the **Graph variable.** Click on the **Labels** button and enter an appropriate title. Click on
OK to view the default histogram. Since you need a class width of 500, edit the graph.
Right-click on the X-axis and select "Edit X scale" from the drop-down menu. On the
Binning tab, select **Cutpoints.** Using the information from the descriptive statistics, you
can see that the minimum value is 1147 and the maximum is 4285. Since you want a
class width of 500, use positions beginning at 1100 and going up to 4600 in steps of 500.
Thus, on the **Binning** tab enter the **Midpoint/cutpoint positions** as **1100:4300/500.**
Click on **OK** to view the changes to the graph.

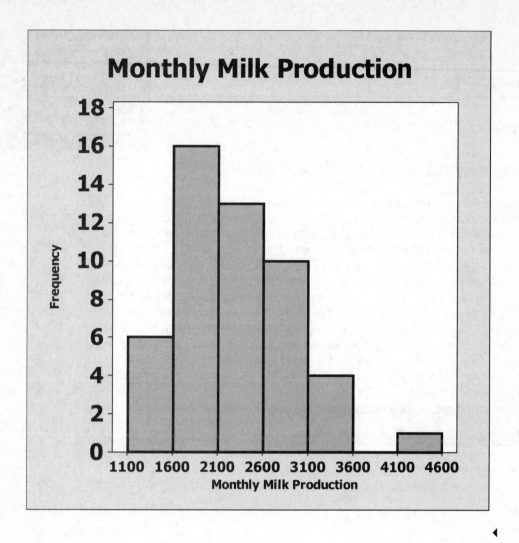

Probability

Section 3.1

▶ Law of Large Numbers (pg. 134) Coin simulation

You can use MINITAB to simulate repeatedly tossing a fair coin and then calculate the empirical probability of tossing a head. This empirical probability will more closely approximate the theoretical probability as the number of tosses gets large. To do this simulation, generate 1000 "tosses" of a fair coin. Let "0" represent a head, and "1" represent a tail. Click on **Calc → Random Data → Bernoulli. Generate** 1000 **rows of data** and **Store in column** C1. Use .5 for the **Probability of Success.** When you click on **OK**, you should see C1 filled with 1's and 0's. To count up the number of "0"s, click on **Stat → Tables → Tally Individual Variables.** Select C1 for **Variables** (by double clicking on C1), and choose both **counts and percents** by clicking on the box to the left of each one. When you click on **OK**, the summary statistics will appear in the Session Window. In the following example, notice that there were 511 heads and 489 tails. Thus, the empirical of tossing a head is .511 (51.1%). This is a very good approximation of the theoretical probability of .5.

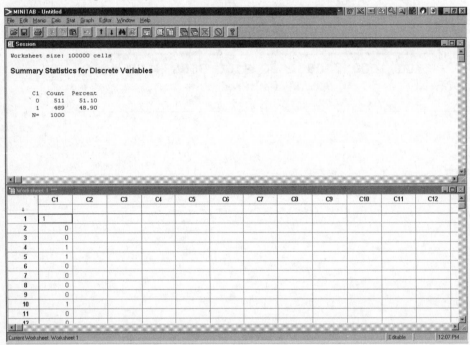

Section 3.2

▶ Exercise 40d (pg. 155) Birthday problem

Simulate the "Birthday Problem" using MINITAB. To do this simulation, the days of the year will be represented by the numbers 1 to 365. Click on **Calc → Random Data → Integer. Generate** 24 **rows of data** (representing the 24 people in the room), **Store In** C1, **Minimum value** is 1 (representing Jan.1st) and **Maximum value** is 365 (representing Dec. 31st). When you click on **OK**, you should see C1 filled with 24 numbers ranging between 1 and 365. These numbers represent the birthdays of the 24 students in the class. The question is: Are there at least two people in the room with the same birthday? To answer this question, we must first summarize the data from this simulation. To do this, click on **Stat → Tables → Tally Individual variables.** On the screen that appears, select C1 for **Variables** and select **Counts.** Next click on **OK**, and in the Session Window, you will see a summary table of C1. This table lists each of the different birthdays that occurred in this simulation, as well as a count of the number of people who had that birthday. Notice that most counts are 1's. If you see a count of "2" or more, then you have at least two people in the room with the same birthday.

Repeat the simulation 9 more times and tally the results each time. How many of the 10 columns had at least two people with the same birthday? The empirical probability that at least two people in a room of 24 people will share a birthday can be calculated as follows: (# of columns having at least two people with the same birthday) divided by 10 (which is the total number of simulations).

◀

▶ Technology Lab (pg. 187)   Composing Mozart variations

3. Click on **Calc → Random Data → Integer. Generate 1 row of data, Store In Column** C1, **Minimum value** is 1 and **Maximum value** is 11. For Part B, repeat these steps but **Generate** 100 **rows of data** (instead of 1 row). To tally the results, click on **Stat → Tables → Tally Individual variables.** Select both **Counts** and **Percents.** The results will appear in the Session Window. Compare the percents to the theoretical probabilities you found in Part A.

5. Click on **Calc → Random Data → Integer. Generate 2 rows of data, Store In** C1, **Minimum value** is 1 and **Maximum value** is 6. Add the two numbers and subtract 1 to obtain the total. For Part B, **Generate** 100 **rows of data, Store In Columns** C1-C2, **Minimum value** is 1 and **Maximum value** is 6. The total will be calculated for each row by adding the two numbers from C1 and C2 and then subtracting 1. To do this, click on **Calc → Calculator. Store result in variable** C3. For **Expression,** type in the following: C1 + C2 - 1. Click on **OK,** and the totals should be in C3. To tally the results, click on **Stat → Tables → Tally Individual variables.** Select both **Counts** and **Percents.** The results will appear in the Session Window. Compare the percents to the theoretical probabilities you found in Part A.

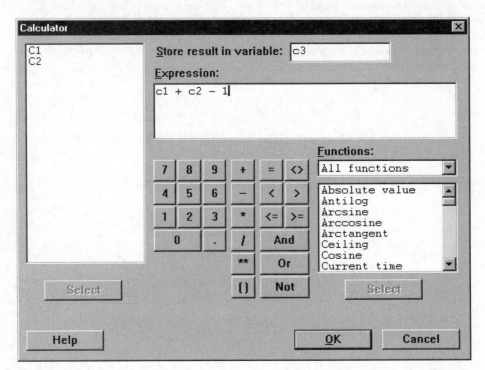

To choose a minuet, Mozart suggested that the player toss a pair of dice 16 times. For the 8th and 16th bars, choose Option 1 if the dice total is odd, and Option 2 if the dice total is even. For each of the other 14 bars, subtract 1 from the dice total. To do this in MINITAB, first simulate rolling the dice 16 times. Click on **Calc → Random Data → Integer. Generate** 16 **rows of data, Store In Columns** C1-C2, **Minimum value** is 1 and **Maximum value** is 6. The total will be calculated for each row by adding the two numbers from C1 and C2 and then subtracting 1. To do this, click on **Calc → Calculator. Store result in variable** C3. For **Expression** type in: C1+C2-1. Click on **OK**, and the totals should be in C3.

The numbers in C3 will be the minuet, except for the 8th and 16th bars. To find these, add C1 + C2 for rows 8 and 16. If the total is odd, choose Option 1 and if the total is even, choose Option 2. For example, the total in row 8 is 3 and Option 1 should be chosen. The total in row 16 is 7, and Option 1 should be chosen again. Thus, the minuet for this simulation is:

8	6	1	6	5	10	6	1
8	10	3	11	9	6	5	1

Notice the 8th and 16th bars are both 1.

Discrete Probability Distributions

Section 4.2

▶ Example 4 (pg. 206) Find a binomial probability

In this example, 67% of American adults consider air conditioning a necessity. A random sample of 100 American adults is selected, so n = 100 and p = .67. Click on **Calc → Probability Distributions → Binomial.** To find the probability that exactly 75 of the 100 adults consider air conditioning a necessity, select **Probability**. This tells MINITAB what type of calculation you want to do. The **Number of Trials** is 100 and the **Probability of Success** is .67. To find the probability of 75 of the 100 consider air conditioning a necessity, click on the circle to the left of **Input Constant** and enter 75 in the box to the right of **Input Constant**. Leave all other fields blank. Click on **OK.**

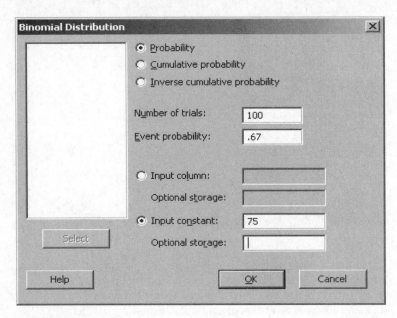

The probability will be displayed in the Session Window. Notice that the probability is .0201004.

Probability Density Function
```
Binomial with n = 100 and p = 0.67

 x  P( X = x )
75   0.0201004
```

◀

> **Example 7 (pg. 209)** Graphing a binomial distribution

In order to graph the binomial distribution, you must first create the distribution and save it in the Data Window. In C1, type in the values of X. Since n=6, the values of X are 0, 1, 2, 3, 4, 5, and 6. Next, use MINITAB to generate the binomial probabilities for n=6 and p=0.6. Click on **Calc → Probability Distributions → Binomial.** Select **Probability.** The **Number of Trials** is 6 and the **Probability of Success** is .6. Select **Input Column** by clicking on the circle on the left. Now, tell MINITAB that the X values are in C1 and that you want the probabilities stored in C2 by entering C1 as the **Input Column** and entering C2 for **Optional Storage.**

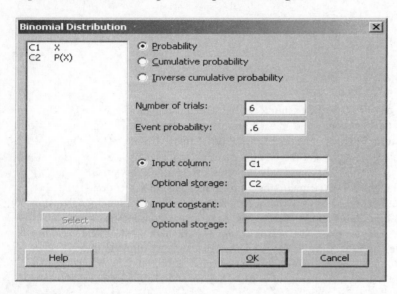

Click on **OK.** The probabilities should now be in C2. Label C1 as "X" and C2 as "P(X)". This will be helpful when you graph the distribution.

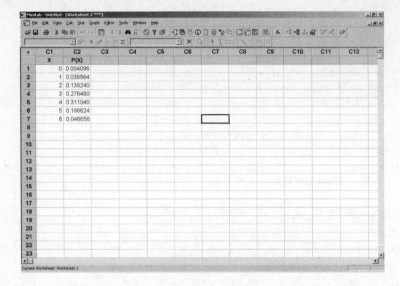

To create the graph, click on **Graph → Bar Chart**. On the screen that appears, below **Bars represent**, click on the down arrow and select **Values from a table**. In the top section, labeled **One column of values**, select the **Simple Bar Chart**. Click on **OK**. On the screen that appears, select 'C2 P(X)' for the **Graph variable** and select 'C1 X' for the **Categorical variable**. Click on **Labels** and enter an appropriate title. Click on OK twice to view the default graph. It is easy to edit the graph and change the label on the X axis and remove the spaces between the bars. On the graph, right click on the label ('X') for the X-axis. Click on **Edit X label** and, in the box for **Text**, enter 'Households.' Next, right click on any numerical value on the X-axis (for example, right click on '1' under the first bar of the chart.) Click on **Edit X scale**. On the screen that appears, click on the check mark to the left of **Gap between clusters**. (This will remove the check mark.) In the box to the right of **Gap between clusters**, enter '0'. Click on OK and the new graph will appear.

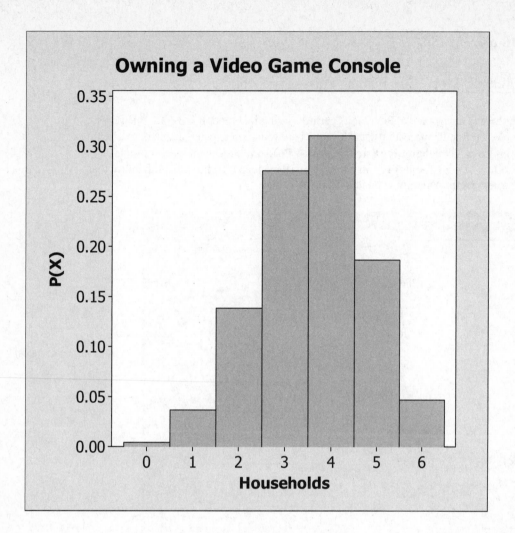

Owning a Video Game Console

Section 4.3

▶ Example 3 (pg. 220) Finding Poisson probabilities

Since there is an average of 3.6 rabbits per acre living in a field, $\mu = 3.6$ for this Poisson example. To find the probability that 7 rabbits are found on any given acre of the field, click on **Calc → Probability Distributions → Poisson.** Since you want a simple probability, select **Probability** and enter 3.6 for the **Mean.** To find the probability that X=7, select **Input constant** and enter 7 for the value.

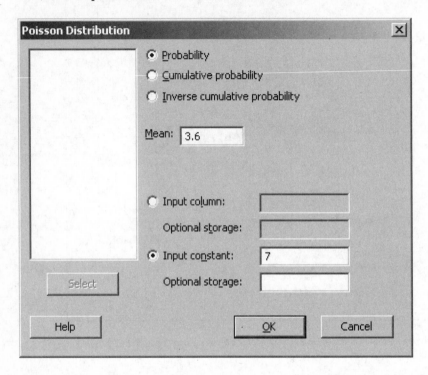

Click on **OK** and the probability will be displayed in the Session Window.

Probability Density Function

Poisson with mean = 3.6

x P(X = x)
7   0.0424841

▶ Technology Lab (pg. 233) Using Poisson distributions as queuing
models

1. First, create the Poisson distribution and save it in the Data Window. In C1, type in the
values of X. Since n=20, the values of X are 0, 1, 2, 3, 4, 5, …20. Next, use MINITAB
to generate the Poisson probabilities for n=20 and μ=4. Click on **Calc → Probability
Distributions → Poisson.** Select **Probability**. The **Mean** is 4. Now, tell MINITAB that
the X values are in C1 and that you want the probabilities stored in C2 by entering C1 as
the **Input Column** and entering C2 for **Optional Storage.** Click **OK.** The probabilities
will be displayed in C2 of the Data Window. Notice, for example, that P(X=4) =
.195367. This probability is the height at X=4 on the histogram that is displayed in the
upper right corner of pg. 237.

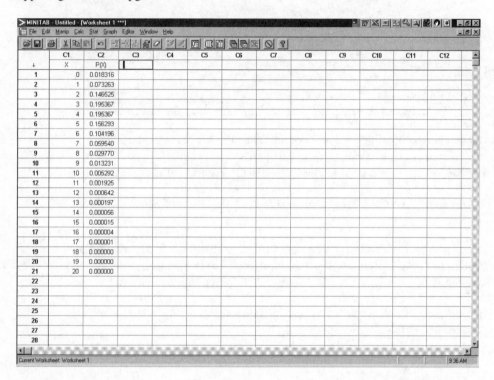

3. To generate 20 random numbers from a Poisson distribution with mean=4, click on
    Calc → Random Data → Poisson. Generate 20 **rows of data** and **Store in
    column** C3. Enter a **Mean** of 4 and click on **OK.**

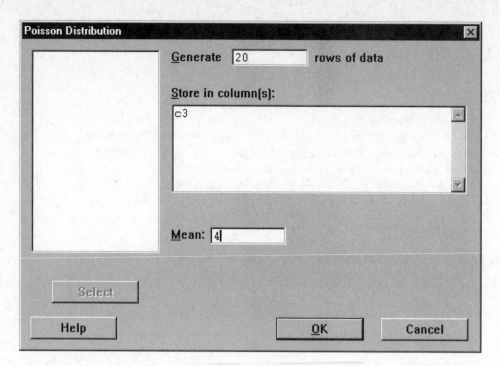

Use the random numbers that are in C3 of the Data Window to create the table of waiting customers and the probabilities. Click on **Stat → Tables → Tally Individual Values.** The **Variable(s)** is C3. Select **Percents**, and click on **OK.**

5. Repeat the steps in Exercise 3, but enter a **Mean** of 5 this time and **Store in column** C4.

6. To calculate P(X=10) for a Poisson random variable with a mean of 5, click on **Calc → Probability Distributions → Poisson.** Since you want a simple probability, select **Probability** and enter 5 for the **Mean** and 10 for the **Input constant.**

7. To find the probabilities for parts a - c, use the Poisson probability distribution that you created in C1 and C2.

Normal Probability Distributions

Section 5.2

▸ Example 3 (pg. 251) Finding normal probabilities

Triglyceride levels in the US are normally distributed with $\mu=134$ and $\sigma=35$. Find the probability that a randomly selected person has a triglyceride level that is less than 80. To do this in MINITAB, click on **Calc → Probability Distributions → Normal.** On the input screen, select **Cumulative probability.** (Cumulative probability *accumulates'* all probability to the left of the input constant.) Enter 134 for the **Mean** and 35 for the **Standard deviation.** Next select **Input Constant** and enter the value 80.

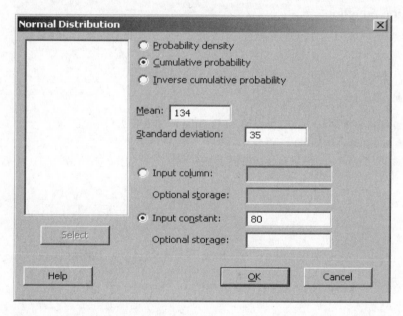

Click on **OK** and the probability should be displayed in the Session Window. As you can see, the probability that a randomly selected person has a triglyceride level that is less than 80 is equal to .0614327.

Cumulative Distribution Function
```
Normal with mean = 134 and standard deviation = 35

 x   P( X <= x )
80    0.0614327
```

▶ Exercise 13 (pg. 253) Height of American men

Heights are normally distributed with $\mu=69.9$ inches and $\sigma=3.0$ inches. To complete parts (a) - (c), you will need MINITAB to give you two probabilities: one using X=66 and the other using X=72. Click on **Calc → Probability Distributions → Normal.** On the input screen, select **Cumulative probability.** Enter 69.9 for the **Mean** and 3.0 for the **Standard deviation.** Next select **Input Constant** and enter the value 66. Click on **OK.** Repeat the above steps using an **Input constant** of 72. Now the Session Window should contain $P(X \le 66)$ and $P(X \le 72)$.

Cumulative Distribution Function
```
Normal with mean = 69.9 and standard deviation = 3

 x   P( X <= x )
66     0.0968005
```

Cumulative Distribution Function
```
Normal with mean = 69.9 and standard deviation = 3

 x   P( X <= x )
72     0.758036
```

So, for Part (a), the $P(X \le 66) = .0968005$. For Part (b), to find the $P(66 \le X \le 72)$, you must subtract the two probabilities. Thus, the $P(66 \le X \le 72) = .758036 - .0968005 = .6612355$. For Part (c), to find $P(X > 72)$, it is $1 - .758036 = .241964$.

◀

Section 5.3

▶ Example 4 (pg. 260) Finding a specific data value

Scores for a California Peace Officer Standards and Training exam are normally distributed with $\mu=50$ and $\sigma=10$. To be eligible for employment, you must score in the top 10%. Find the lowest score you can earn and still be eligible for employment. To do this in MINITAB, click on **Calc → Probability Distributions → Normal.** On the input screen, select **Inverse Cumulative probability.** Enter 50 for the **Mean** and 10 for the **Standard deviation.** For this type of problem, the **Input constant** will be the area to the left of the X-value we are looking for. This input constant will be a decimal number between 0 and 1. For this example, select **Input Constant** and enter the value .90 since 10% of the test scores are above this number and therefore, 90% are below this number. Click on **OK** and the X-value should be in the Session Window. Notice that the test score that qualifies you for employment is 62.8155.

Inverse Cumulative Distribution Function

```
Normal with mean = 50 and standard deviation = 10

P( X <= x )        x
       0.9  62.8155
```

◀

Section 5.4

▶ Example 6 (pg. 273) Finding Probabilities for x and $\bar{x}$

Credit card balances are normally distributed, with a mean of $3173 and a standard deviation of $1120.

1. To find the probability that a randomly selected card holder has a balance less than $2700, click on **Calc → Probability Distributions → Normal.** On the input screen, select **Cumulative probability.** Enter 3173 for the **Mean** and 1120 for the **Standard deviation.** Next select **Input Constant** and enter the value 2700. Click on **OK** and the probability should appear in the Session Window.

    Cumulative Distribution Function
    ```
    Normal with mean = 3173 and standard deviation = 1120

      x   P( X <= x )
    2700     0.336395
    ```

2. To find the probability that the *mean* balance of 25 card holders is less than $2700, you will need to calculate the standard deviation of $\bar{x}$ which is equal to $1120/\sqrt{25}$ = 224. (Use a hand calculator for this calculation.) Now let MINITAB do the rest for you. Click on **Calc → Probability Distributions → Normal.** On the input screen, select **Cumulative probability.** Enter 3173 for the **Mean** and 224 for the **Standard deviation.** Next select **Input Constant** and enter the value 2700. Click on **OK** and the probability should appear in the Session Window.

    Cumulative Distribution Function
    ```
    Normal with mean = 3173 and standard deviation = 224

    x   P( X <= x )
    2700    0.0173601
    ```

◀

▸ Technology Lab (pg. 299) Age distribution in the United States

In this lab, you will compare the age distribution in the United States to the sampling distribution that is created by taking 36 random samples of size n=40 from the population and calculating the sample means.

Open worksheet **Tech5** which is found in the **ch05** folder. C1 should now contain the mean ages from the 36 random samples.

1. From the table on page 299 in your textbook, enter the Class Midpoints into C2. To enter the midpoints, click on **Calc → Make Patterned Data → Simple Set of Numbers.** On the input screen, you should **Store patterned data in** C2, **from the first value** of 2 **to the last value** of 97, **in steps of** 5. Click on **OK** and the midpoints should now be in C2. Next enter the relative frequencies, converted to proportions, into C3. So for a relative frequency of 6.9%, you should enter .069 into C3. The mean of this distribution is $\Sigma x\ p(x)$. To calculate the mean, you will have to multiply C2 and C3. To do this, click on **Calc → Calculator.** Type in the **Expression** C2 * C3 and **store result in variable** C4. Click on **OK** and in C4 you should now see the product of C2 and C3.

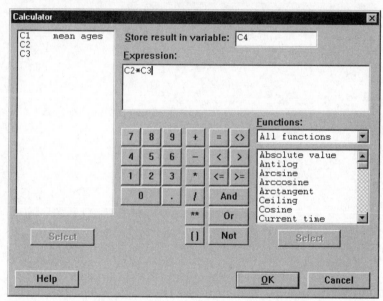

  Now find the sum of C4. Click on **Calc → Column Statistics.** On the input screen, select **Sum** and use C4 for the **Input variable.** Click on **OK** and the column sum should be in the Session Window. As you can see, the mean age in the United States is 37.055.

2. The 36 sample means are in C1. To find the mean, click on **Stat → Basic Statistics → Display Descriptive Statistics.** Select C1 for the **Variable** and click on **OK.** The descriptive statistics will be displayed in the Session Window. The mean of the

set of sample means is 36.209 and the standard deviation is 3.552. (You will need the standard deviation for question 6.)

4. To draw the histogram, click on **Graph → Histogram.** Select C1 for the **Graph variable.** In order to create a *relative frequency* histogram, click on **Options** and select **Percent.** Click on **OK** twice and you should be able to view the histogram.

5. To find the standard deviation of the ages of Americans, you must use the formula for the standard deviation of a Discrete Random variable, found on page 195 in the textbook. The shortcut formula will make this calculation easier. Use the formula $\Sigma x^2 \, p(x) - \mu^2$ and take the square root of this value. In MINITAB, first square all the midpoints. Click on **Calc → Calculator.** Type in the **Expression** C2 * C2 and **store result in variable** C5. Click on **OK** and in C5 you should now see the midpoints squared. To calculate $x^2 \, p(x)$, you must multiply C5 by C3. Click on **Calc → Calculator.** Type in the **Expression** C5 * C3 and **store result in variable** C6.

	C1	C2	C3	C4	C5	C6	C7	C8	C9	C10	C11	C12
	Mean ages	Midpoints	P(X)	X*P(X)	X*X	X*X*P(X)	StDev					
1	28.14	2	0.069	0.138	4	0.276	22.8105					
2	31.56	7	0.066	0.462	49	3.234						
3	36.86	12	0.066	0.792	144	9.504						
4	32.37	17	0.071	1.207	289	20.519						
5	36.12	22	0.069	1.518	484	33.396						
6	39.53	27	0.070	1.890	729	51.030						
7	36.19	32	0.064	2.048	1024	65.536						
8	39.02	37	0.069	2.553	1369	94.461						
9	35.62	42	0.071	2.982	1764	125.244						
10	36.30	47	0.075	3.525	2209	165.675						
11	34.38	52	0.071	3.692	2704	191.984						
12	32.98	57	0.061	3.477	3249	198.189						
13	36.41	62	0.050	3.100	3844	192.200						
14	30.24	67	0.037	2.479	4489	166.093						
15	34.19	72	0.029	2.088	5184	150.336						
16	44.72	77	0.024	1.848	5929	142.296						
17	38.84	82	0.019	1.558	6724	127.756						
18	42.87	87	0.012	1.044	7569	90.828						
19	38.90	92	0.005	0.460	8464	42.320						
20	34.71	97	0.002	0.194	9409	18.818						
21	34.13											
22	38.25											
23	38.04											
24	34.07											
25	39.74											
26	40.91											
27	42.63											

Now find the sum of C6. Click on **Calc → Column Statistics.** On the input screen, select **Sum** and use C6 for the **Input variable.** Click on **OK** and the column sum should be in the Session Window. As you can see, $\Sigma x^2 \, p(x)$ is 1889.69. Next, subtract μ^2 from 1889.69 and take the square root. (Recall that $\mu = 37.055$) Click on **Calc → Calculator.** Type in the **Expression** SQRT(1889.69-37.055*37.055) and **store result in variable** C7. The standard deviation is the number now in C7, 22.8105.

6. The standard deviation of the 36 sample means can be found in the descriptive statistics that you produced for question 2.

Confidence Intervals

Section 6.1

▶ Example 4 (pg. 308)   Construct a 99% confidence
                        interval

Enter the data for Example 1 (on page 304 of text) into C1. First find the standard
deviation of the data. Click on **Calc → Column Statistics.** Select **Standard deviation**
and enter C1 for the **Input variable.** Click on **OK** and the standard deviation will be
displayed in the Session Window (s=52.63). To construct the confidence interval, click
on **Stat → Basic Statistics → 1-Sample Z.** Enter C1 in **Samples in columns**. For
Standard deviation, enter the rounded sample standard deviation of 53.

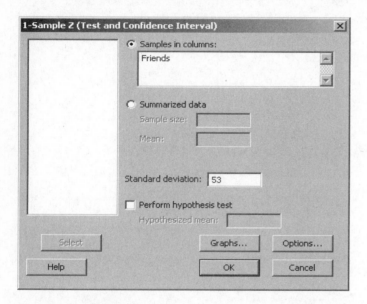

Next, click on **Options** and enter 99.0 for the **Confidence Level.**

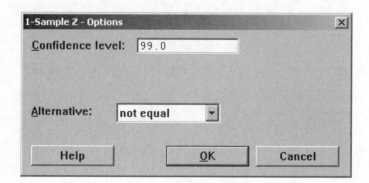

Click on **OK** twice and the results will be displayed in the Session Window.
As you can see, the 99% confidence interval is (109.21, 152.39).

One-Sample Z: Friends

The assumed standard deviation = 53

Variable	N	Mean	StDev	SE Mean	99% CI
Friends	40	130.80	52.63	8.38	(109.21, 152.39)

▶ Exercise 49 (pg. 313) Construct 90% and 99% confidence intervals for mean DVR watching time

Enter the data into C1. To construct the confidence interval, click on **Stat → Basic Statistics → 1-Sample Z.** Enter C1 for **Samples in columns.** Enter the assumed value of 1.3 for the **Standard deviation.** Next, select **Options** and enter 90.0 for the **Confidence Level.** Click on **OK** and the interval will be displayed in the Session Window. Repeat the above steps using 99.0 for the **Confidence Level.**

One-Sample Z: Time

The assumed standard deviation = 1.3

Variable	N	Mean	StDev	SE Mean	90% CI
Time	20	15.100	5.875	0.291	(14.622, 15.578)

One-Sample Z: Time

The assumed standard deviation = 1.3

Variable	N	Mean	StDev	SE Mean	99% CI
Time	20	15.100	5.875	0.291	(14.351, 15.849)

▶ Exercise 63 (pg. 315) Construct a 95% confidence interval for
airfare prices

Open worksheet **Ex6_1-63.mtp** in the **ch06** Folder. The airfares are in C1. Find the
standard deviation of the sample. Click on **Calc → Column Statistics.** Select **Standard
deviation** and enter C1 for the **Input variable.** Click on **OK** and the standard deviation
will be displayed in the Session Window. To construct the confidence interval, click on
Stat → Basic Statistics → 1-Sample Z. Enter C1 in **Samples in columns.** For
Standard deviation, enter the standard deviation that is displayed in the Session
Window. Next, select **Options** and enter 95.0 for the **Confidence Level.** Click **OK**
twice.

One-Sample Z: Boston to Chicago airfares

The assumed standard deviation = 12.65

Variable	N	Mean	StDev	SE Mean	95% CI
Boston to Chicago	32	217.13	12.65	2.24	(212.74, 221.51)

◀

Section 6.2

Example 2 (pg. 320) Construct a 95% confidence interval for
the mean temperature of coffee sold

Using the summarized data found on page 327 of the text, construct a 95% confidence
interval for the temperatures of the coffee sold at 16 randomly selected restaurants.
Since n=16 and the population standard deviation is unknown, you should construct a t-
interval for this problem. Click on **Stat → Basic Statistics → 1-Sample t.** Select
Summarized data. The **Sample size** is 16, the **Mean** is 162, and the **Standard
deviation** is 10.   Next, select **Options** and enter 95.0 for the **Confidence Level.** Click on
OK twice and the output will be displayed in the Session Window.

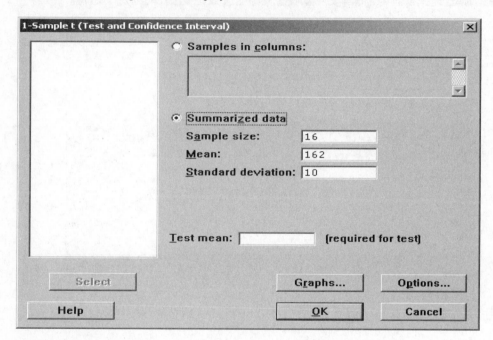

The following confidence interval will be displayed in the Session Window.

One-Sample T

```
 N    Mean   StDev  SE Mean      95% CI
16  162.000  10.000   2.500  (156.671, 167.329)
```

▶ Exercise 25 (pg. 324)   Construct a 99% confidence interval for
                          the mean SAT scores

Enter the SAT scores into C1. Since n=12 and the population standard deviation is
unknown, you should construct a t-interval for this problem. Click on **Stat → Basic
Statistics → 1-Sample t.** Enter C1 in **Samples in column**. Next, select **Options** and
enter 99.0 for the **Confidence Level.** Click on **OK** twice and the output will be displayed
in the Session Window.

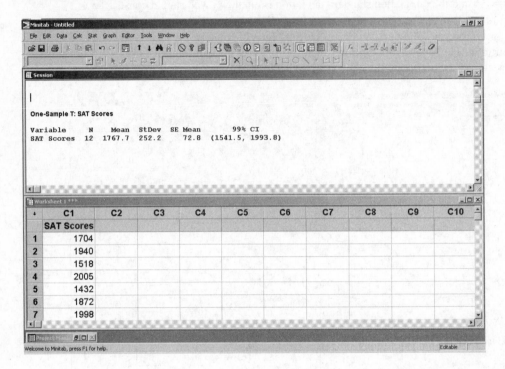

Section 6.3

▶ **Example 2 (pg. 329)** Construct a 95% confidence interval for *p*

From Example 1, on page 327 of the textbook, you know that 662 of 1000 American adults said that it is acceptable to check their personal email while at work. To construct a 95% confidence interval, click on **Stat → Basic Statistics → 1 Proportion.** Select **Summarized Data.** The **Number of trials** is 1000 and the **Number of events** is 662

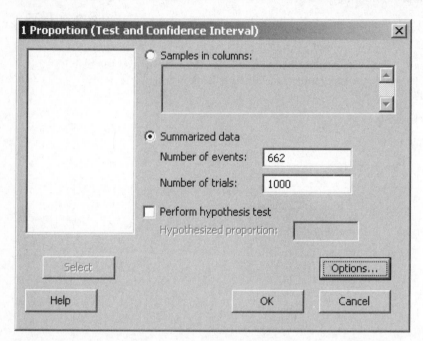

Next, to select the confidence level, click on **Options.** Enter 95.0 for the **Confidence Level.** Also, select **Use test and interval based on normal distribution.**

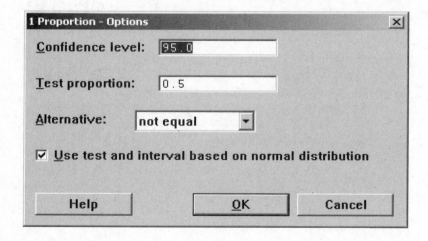

Click on **OK** twice and the output will be displayed in the Session Window.

Test and CI for One Proportion

Sample   X   N Sample p   95% CI
1   662 1000 0.662000 (0.632682, 0.691318)

Using the normal approximation.

Notice the interval is (.63, .69). This means that with 95% confidence, you can say that the proportion of adults who say checking personal email at work is acceptable is between 63% and 69%.

◀

▸ Exercise 13 (pg. 332) Going green

A survey of 3110 adults found that 1435 started paying their bills online in the last year. Construct a 95% confidence interval for the true proportion of adults who pay their bills online. Click on **Stat → Basic Statistics → 1 Proportion.** Select **Summarized Data.** The **Number of events** is 1435 and the **Number of trials** is 3110. Next, to select the confidence level, click on **Options.** Enter 95.0 for the **Confidence Level.** Also, select **Use test and interval based on normal distribution.** Click on **OK** twice and the results will be in the Session Window.

Test and CI for One Proportion

Sample   X   N   Sample p   95% CI
1   1435 3110 0.461415 (0.443895, 0.478935)

Using the normal approximation.

◂

▶ Technology Lab (pg. 351)   The Gallup Organization

1. Click on **Stat → Basic Statistics → 1 Proportion.** Select **Summarized Data.** The **Number of trials** is 1025 and the **Number of events** is 308 (.30 x 1025=307.5). Next, to select the confidence level, click on **Options.** Enter 95.0 for the **Confidence Level.** Also, select **Use test and interval based on normal distribution.** Click on **OK** twice and the results will be in the Session Window.

2. Hilary Clinton was named by 16% of the people in the sample. That means that 164 (.16 x 1025 = 164) of the 1025 named Hilary Clinton. Use the steps in question 1 to construct the 95% confidence interval. This time the **Number of events** is 164.

4. Sarah Palin was named by 15% of the people in the sample. That means that 154 (.15 x 1025 = 153.75) of the 1025 named Sarah Palin. Use the steps in question 1 to construct the 95% confidence interval. This time the **Number of events** is 154.

5. To do this simulation, you will generate random binomial data with n=1025 and p=.18. The result displayed in each cell of C1 will represent the number of people who named Sarah. Click on **Calc → Random Data → Binomial. Generate** 200 **rows of data** and **Store in column** C1. The **Number of trials** is 1025 and the **Probability of success** is .18. When you click on **OK**, C1 will contain a simulation of 200 samples.

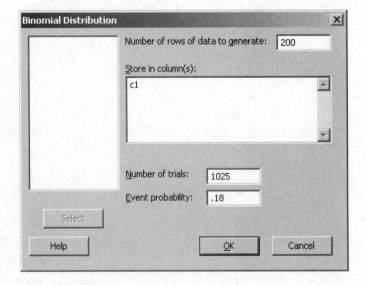

To calculate the sample proportions, click on **Calc → Calculator**. Enter C1/1025 for the **Expression** and **Store the result in** C2. Click on **OK**. Now C2 contains 200 values of the sample proportion. Sort C2 to find the smallest and largest value. Click on **Data → Sort. Sort column** C2, **By column** C2 and **Store sorted data in column of current worksheet** C3.

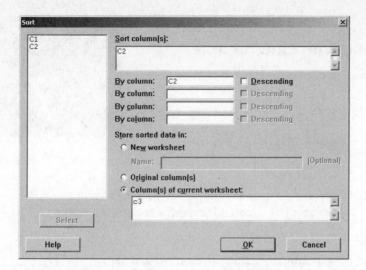

Click on **OK** and C3 should contain the sample proportions sorted from smallest to largest. Thus, the smallest value should be in Row 1 and the largest value should be in Row 200. Since this is random data, results will vary each time this is repeated.

Hypothesis Testing with One Sample

CHAPTER 7

Section 7.2

▸ Example 5 (pg. 375) Hypothesis testing using *P*-values

You think that the average cost of $22500 for bariatric surgery is incorrect, so you randomly select 30 patients and determine the mean cost for surgery is $21,454 with a standard deviation of $3,015. Is there enough evidence to support your claim at $\alpha = .05$? Use the *P*-value to interpret.

Click on **Stat → Basic Statistics → 1-Sample Z.** Click on **Summarized data.** Enter 30 for **Sample size,** 21454 for **Mean,** 3015 for **Standard deviation**, and 22500 for **Test mean.**

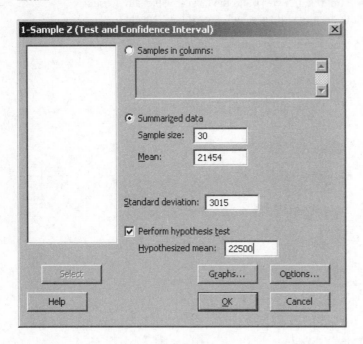

Since the claim is "the mean is different from $22,500", you will perform a two-tailed test. Click on **Options**, and set **Alternative** to "not equal".

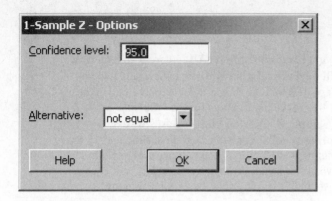

Click on **OK** twice and the results should be displayed in the Session Window.

One-Sample Z

```
Test of mu = 22500 vs not = 22500
The assumed standard deviation = 3015

N    Mean  SE Mean       95% CI        Z      P
30   21545     550   (20466, 22624)  -1.73  0.083
```

Notice that both the test statistic and the *P*-value are given. From the output, note that $z = -1.73$ and $p = .083$. Since the *P*-value is larger than α, you would fail to reject the null hypothesis.

◀

▶ Exercise 33 (pg. 383) Years taken to quit smoking permanently

Open worksheet **Ex7_2-33,** which is found in the **ch07** MINITAB folder. Click on **Calc → Column Statistics.** Select **Standard deviation** for the **Statistic** to be calculated and enter C1 for the **Input Variable.** Click on **OK** and the standard deviation will be in the Session Window. You will enter this value in the input screen for the 1-Sample Z test. Click on **Stat → Basic Statistics → 1-Sample Z.** Click on **Stat → Basic Statistics → 1-Sample Z.** Enter C1 for **Samples in columns,** 4.288 for **Standard deviation,** and 15 for **Test mean.** Since the claim is "the mean time is 15 years", you will perform a two-tailed test. Click on **Options** and use the down arrow beside **Alternative** to select "not equal". Click on **OK** and the results of the test should be displayed in the Session Window.

One-Sample Z: Years

```
Test of mu = 15 vs not = 15
The assumed standard deviation = 4.288


Variable    N    Mean  StDev  SE Mean      95% CI         Z      P
Years      32  14.834  4.288    0.758  13.349, 16.320) -0.22  0.827
```

Since $p=0.827$ and is larger than 0.05, you fail to reject the null hypothesis.

◀

▶ Exercise 39 (pg. 384) Nitrogen dioxide levels

Open worksheet **Ex7_2-39,** which is found in the **ch07** MINITAB folder.   Click on **Calc**
→ **Column Statistics.** Select **Standard deviation** for the **Statistic** to be calculated and
enter C1 for the **Input Variable.** Click on **OK** and the standard deviation will be in the
Session Window. You will enter this value in the input screen for the 1-Sample Z test.
Click on **Stat** → **Basic Statistics** → **1-Sample Z.** Enter C1 for **Samples in columns,**
9.164 for **Standard deviation**, and 32 for **Test mean.** Since the claim is "the mean is
greater than 32 parts per billion", you will perform a right-tailed test. Click **Options** and
then on the down arrow beside **Alternative** and select "greater than". Click on **OK** twice
and the results of the test should be displayed in the Session Window.

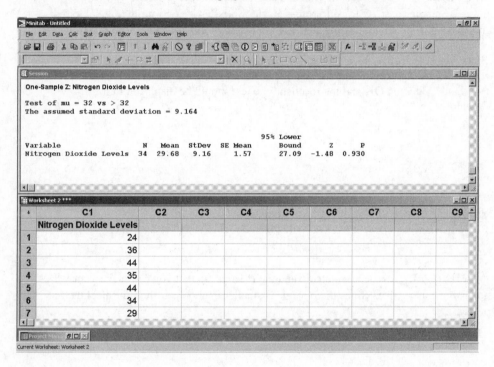

Since the *P*-value (.93) is greater than the level of significance α = .05, the data fail to
support the scientist's estimate.

◀

Section 7.3

▶ Example 4 (pg. 390) Testing μ with a small sample

A used car dealer says that the mean price of a 2008 Honda CR-V is at least $20,500. You suspect this claim is incorrect. At $\alpha = .05$, is there enough evidence to reject the dealer's claim?

Based on a random sample of 14 similar vehicles, you found that the mean price is $19,850 with a standard deviation of $1084. Be sure to enter the data in numeric form. So enter the amount $19,850 as 19850. Since this is a small sample problem, you will be performing a 1-Sample t-test. Click on **Stat** → **Basic Statistics** → **1-Sample t.** Click on **Summarized data,** and enter a **Sample size** of 14, a **Mean** of 19850, and a **Standard deviation** of 1084. Click on **Perform hypothesis test** and enter 20500 for **Hypothesized mean.** Since you suspect that the used car dealer's claim is too high, you will perform a left-tailed test. Click on **Options** and then on the down arrow beside **Alternative** to select "less than". Click on **OK** and the results of the test should be displayed in the Session Window.

```
One-Sample T

Test of mu = 20500 vs < 20500

                               95% Upper
  N    Mean   StDev   SE Mean    Bound       T      P
 14   19850    1084       290    20363   -2.24  0.021
```

Notice that MINITAB gives the test statistic and the P-value, so that you can make your conclusion using either value. Since the P-value is smaller than α, you should Reject the null hypothesis.

◀

▶ Exercise 21 (pg. 394) Amount of waste recycled

You want to test the claim that the mean amount of waste recycled is more than 1 pound per person per day. In a random sample of 13 adults, the mean amount recycled is 1.5 lbs with a standard deviation of 0.28 lbs. Use $\alpha = 0.10$. Click on **Stat → Basic Statistics →1-Sample t.** Click on **Summarized data,** and enter a **Sample size** of 13, a **Mean** of 1.5, and a **Standard deviation** of 0.28. Click on **Perform hypothesis test** and enter 1 for **Hypothesized mean**. Since the claim is "adults recycle more than 1 lb", you will perform a right-tailed test. Click on **Options** and then on the down arrow beside **Alternative** to select "greater than". Click on **OK** twice and the results of the test should be displayed in the Session Window.

One-Sample T

Test of mu = 1 vs > 1

				95% Lower		
N	Mean	StDev	SE Mean	Bound	T	P
13	1.5000	0.2800	0.0777	1.3616	6.44	0.000

Since $p=0.00$, you would reject the null hypothesis. The data support the claim that adults recycle more than 1 lb per person per day.

◀

▶ Exercise 29 (pg. 395) Class size

Enter the data into C1 of a Minitab worksheet. Click on **Stat → Basic Statistics →1-Sample t.** Enter C1 for **Samples in columns,** click on **Perform hypothesis test** and the **Hypothesized mean** is 32. Since the claim is "mean class size is fewer than 32 students", you will perform a left-tailed test. Click on **Options** and then on the down arrow beside **Alternative** to select "less than". Click on **OK** and the results of the test should be displayed in the Session Window.

One-Sample T: Class size

```
Test of mu = 32 vs < 32


                                           95% Upper
Variable      N    Mean   StDev  SE Mean   Bound       T     P
Class size   18   30.167  4.004   0.944    31.808   -1.94  .034
```

Since $p=0.034$, you should reject the null hypothesis. There is sufficient evidence to show that full-time faculty have fewer than 32 students per class.

◀

Section 7.4

▸ Example 2 (pg. 400)   Hypothesis test for a proportion

Of 200 college graduates, 21% say that a college degree is not worth the cost. At $\alpha = .10$, is there enough evidence to reject the claim that 25% of college graduates believe that the degree is not worth the cost?

Click on **Stat → Basic Statistics → 1-Proportion.** The data is given in a summarized form, so select **Summarized data.** Since 21% of the sample were in favor, the **Number of events** is 42 (.21 * 200). Enter 200 for the **Number of trials.** Click on **Perform hypothesis test** and enter .25 for **Hypothesized proportion.**

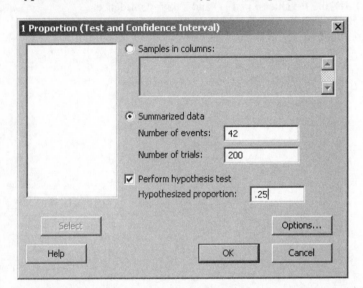

Click on **Options** and select "not equal" for the **Alternative.** Since np and nq are both larger than 5, click on **Use test and interval based on normal distribution,** and then click on **OK** twice.

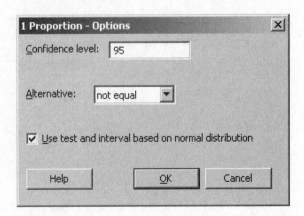

The results should be displayed in the Session Window.

Test and CI for One Proportion

Test of p = 0.25 vs p not = 0.25

Sample   X   N   Sample p   95% CI   Z-Value P-Value
1   42 200 0.210000 (0.154, 0.266)   -1.31   0.191

Using the normal approximation.

Notice that the test statistic ($z = -1.31$), the *P*-value ($p = .191$) and a 95% confidence interval for the true proportion of college graduates who say the college degree is not worth the cost are all displayed in the output. With such a large *P*-value, you should fail to reject the null hypothesis.

◀

▶ Exercise 9 (pg. 401) Smokers

A medical researcher says that less than 25% adults are smokers. In a random sample of 200 adults, 18.5% are smokers. Is there enough evidence to reject the researcher's claim?

Click on **Stat → Basic Statistics → 1-Proportion.** The data is given in a summarized form, so select **Summarized data.** Enter 200 for the **Number of trials.** Since 18.5% of the sample are smokers, the **Number of events** is 37 (.185 * 200 = 37). Click on **Perform hypothesis test** and enter .25 for **Hypothesized proportion.** Click on **Options** and select "less than" for the **Alternative.** Click on **Use test and interval based on normal distribution,** and then click on **OK** twice.

Test and CI for One Proportion

Test of p = 0.25 vs p < 0.25

```
                          95% Lower
Sample   X    N   Sample p    Bound   Z-Value  P-Value
1        37  200  0.185000  0.139838   -2.12    0.017
```

Since $p=0.017$ and is less than the significance level 0.05, you would reject the null hypothesis.

▸ Exercise 11 (pg. 401) Cell phones and driving

A research center claims that at most 50% of people believe that drivers should be
allowed to use cell phones with hands-free devices. In a random sample of 150 people,
58% believe cell phone use with hands-free devices should be allowed. Is there enough
evidence to reject the center's claim at a significance level of 0.01?

Click on **Stat → Basic Statistics → 1-Proportion.** The data is given in a summarized
form, so select **Summarized data**. The **Number of events** is 87 (.58 x 150). Enter 150
for the **Number of trials.**   Click on **Perform hypothesis test** and enter .50 for
Hypothesized proportion. Click on **Options** and select "greater than" for the
Alternative. Click on **Use test and interval based on normal distribution,** and then
click on **OK** twice.

Test and CI for One Proportion

Test of p = 0.5 vs p > 0.5

				95% Lower		
Sample	X	N	Sample p	Bound	Z-Value	P-Value
1	87	150	0.580000	0.513714	1.96	0.025

Using the normal approximation.

Since $p=0.025$ is more than the significance level 0.01, you would fail to reject
the null hypothesis.

◀

Section 7.5

▶ Example 4 (pg. 407)   Hypothesis test for a variance

Click on **Stat → Basic Statistics → 1-Variance.** First select **Enter Variance** in the top drop-down since this problem is given in terms of the variance. The data is given in a summarized form, so select **Summarized data**. Enter 41 for the **Sample size.** Enter .27 for the **Sample variance**. Click on **Perform hypothesis test** and enter .25 for **Hypothesized variance.** Click on **Options** and select "greater than" for the **Alternative** and then click on **OK** twice.

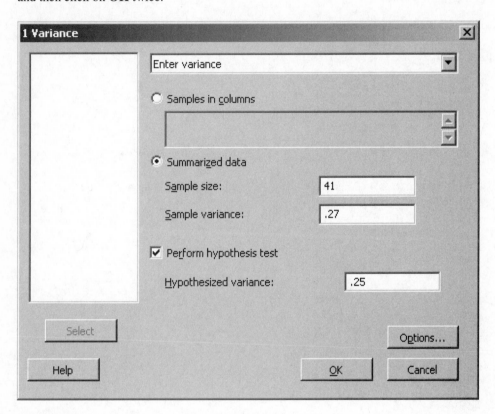

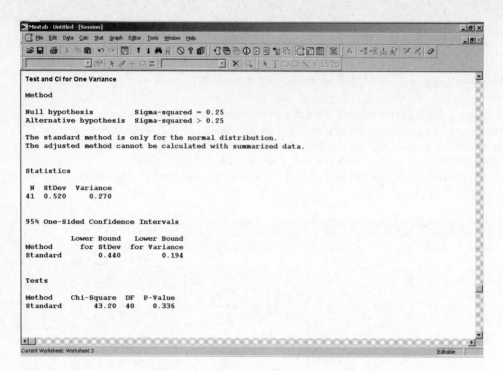

The chi-square test statistic is 43.2 and the *P*-value is 0.336. With such a large *P*-value, you would fail to reject the null hypothesis.

▸ Example 5 (pg. 408)   Hypothesis test for a standard deviation

Click on **Stat → Basic Statistics → 1-Variance.** First select **Enter Standard deviation** in the top drop-down since this problem is given in terms of the standard deviation. The data is given in a summarized form, so select **Summarized data**. Enter 25 for the **Sample size.** Enter 1.1 for the **Sample standard deviation.** Click on **Perform hypothesis test** and enter 1.4 for **Hypothesized variance.** Click on **Options** and select "less than" for the **Alternative** and then click on **OK** twice.

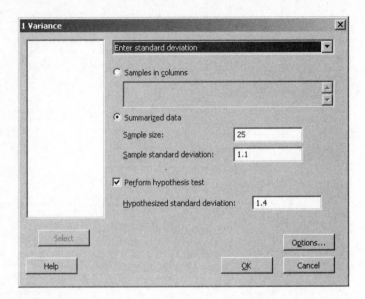

Test and CI for One Standard Deviation

Method

Null hypothesis   Sigma = 1.4
Alternative hypothesis   Sigma = < 1.4

The standard method is only for the normal distribution.
The adjusted method cannot be calculated with summarized
data.

Statistics

 N   StDev   Variance
25   1.10   1.21

95% One-Sided Confidence Intervals

        Upper
        Bound
          for   Upper Bound

```
Method     StDev  for Variance
Standard   1.45            2.10


Tests

Method     Chi-Square  DF  P-Value
Standard        14.82  24    0.074
```

The Chi-square test statistic is 14.82 and the *P*-value is 0.074. Since the *P*-value is greater than 0.05, you would fail to reject the null hypothesis.

▶ Technology Lab (pg. 423) The case of the vanishing women

4. Click on **Stat → Basic Statistics → 1-Proportion.** The data is given in a
    summarized form, so select **Summarized data**. Enter 100 for the **Number of trials.**
    Since 9 women were selected, the **Number of events** is 9. Click on **Options**. Enter
    .2914 for the **Test Proportion** because 29.14% of the original sample were women,
    and select "not equal" for the **Alternative**. Click on **Use test and interval based on
    normal distribution,** and then click on **OK** twice.

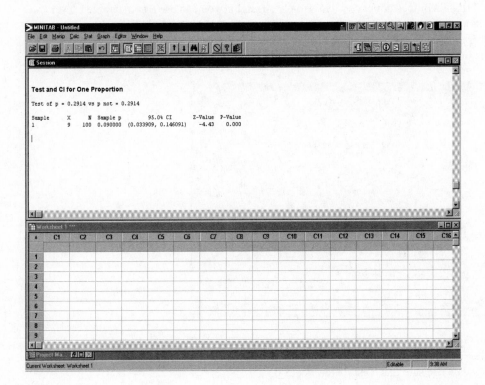

Section 8.1

▶ Exercise 31 (pg. 437) Watching more TV?

Open worksheet **Ex8_1-31.mtp**, which is found in the **ch08** MINITAB folder. The 1981 data (Time A) is in C1 and the new data (Time B) is in C2. Notice that for both samples, $n = 30$. MINITAB does not have a two-sample z-test, but you can use a two-sample t-test instead since the t distribution becomes very similar to the normal distribution as the sample size approaches 30. Click on **Stat → Basic Statistics → 2-Sample t.** Select **Samples in different columns** and enter C1 for the **First** and C2 for the **Second** column.

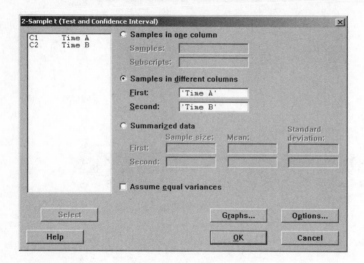

Click on **Options**, and then on the down arrow beside **Alternative** and select **greater than** since the sociologist claims that children spent more time watching TV in 1981 than they do today. Be sure that **Test difference** is 0.

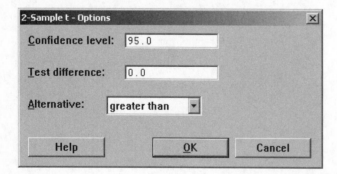

Click on **OK** twice and the results of the test should be displayed in the Session Window.
Two-Sample T-Test and CI: Time A, Time B

```
Two-sample T for Time A vs Time B

          N    Mean   StDev   SE Mean
Time A   30   2.130   0.490    0.089
Time B   30   1.757   0.470    0.086


Difference = mu (Time A) - mu (Time B)
Estimate for difference:  0.373
95% lower bound for difference:  0.166
T-Test of difference = 0 (vs >): T-Value = 3.01  P-Value =
0.002  DF = 57
```

Notice that the test statistic is T = 3.01 with a *P*-value = 0.002. Since this *P*-value is so small, you would Reject H_0 at any α level. Thus, the sociologist's claim is true – children watched more TV in 1981.

◀

▸ Exercise 33 (pg. 438) Difference between washer diameters

Open worksheet **Ex8_1-33.mtp**, which is found in the **ch08** MINITAB folder. The
diameters from the first method are in C1 and the diameters from the second method are
in C2. Notice that for both samples, n = 35. MINITAB does not have a two-sample z-
test, but you can use a two-sample t-test instead since the t distribution becomes very
similar to the normal distribution as the sample size gets larger than 30. Click on **Stat →
Basic Statistics → 2-Sample t**. Select **Samples in different columns** and enter C1 for
the **First** and C2 for the **Second** column. Click on **Options**, and then on the down arrow
beside **Alternative** and select **not equal** since the production engineer claims there is no
difference between the two methods. Click on **OK** twice and the results of the test should
be displayed in the Session Window.

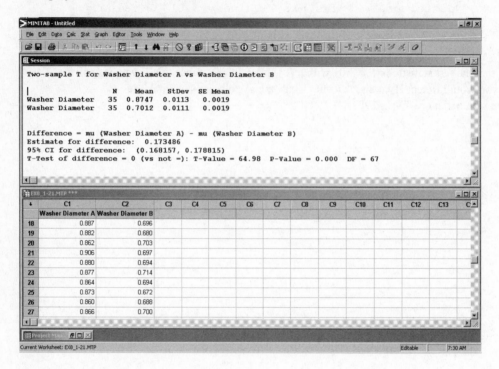

Since the *P*-value is so small, you would Reject H_0 at any α level.

▶ Exercise 34 (pg. 438)   Difference between nut diameters

Open worksheet **Ex8_1-34.mtp**, which is found in the **ch08** MINITAB folder. The diameters from the first method are in C1 and the diameters from the second method are in C2. Notice that for both samples, n = 40. MINITAB does not have a two-sample z-test, but you can use a two-sample t-test instead since the t distribution becomes very similar to the normal distribution as the sample size gets larger than 30. Click on **Stat → Basic Statistics → 2-Sample t.** Select **Samples in different columns** and enter C1 for the **First** and C2 for the **Second** column. Click on **Options**, and then on the down arrow beside **Alternative** and select **not equal** since the production engineer claims there is no difference between the two methods. Click on **OK** twice and the results of the test should be displayed in the Session Window.

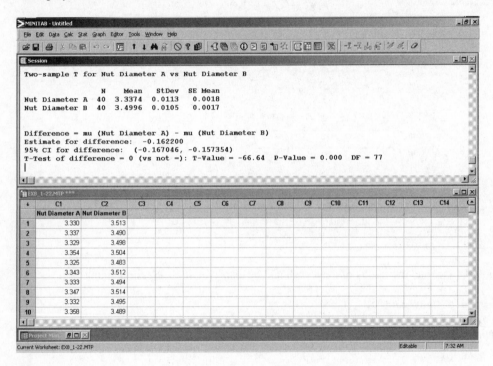

Since the P-value is so small, you would Reject H_0 at any α level.

Section 8.2

▸ Example 1 (pg. 444) Math test scores

The results of a state math test for random samples of students taught by 2 different teachers at the same school is shown on the left of page 444 in the text. Click on **Stat →Basic Statistics → 2-Sample t.** Select **Summarized data** and enter the Teacher 1 data for the **First** sample and the Teacher 2 data for the **Second** sample. Click on **Options**, and then on the down arrow beside **Alternative** and select **not equal** since you want to test whether the mean test scores are different.

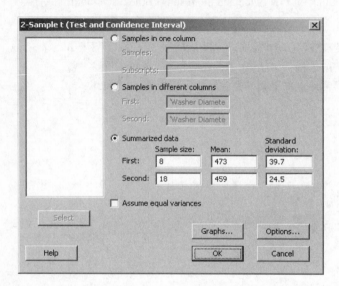

Click on **OK** twice and the results of the test should be displayed in the Session Window.

```
Two-Sample T-Test and CI

Sample   N   Mean   StDev   SE Mean
1        8   473.0   39.7        14
2       18   459.0   24.5       5.8


Difference = mu (1) - mu (2)
Estimate for difference:  14.0
95% CI for difference:  (-20.3, 48.3)
T-Test of difference = 0 (vs not =): T-Value = 0.92
P-Value = 0.380   DF = 9
```

Notice that the test statistic is T= 0.92. Since the *P*-value = 0.38 and is greater than the α-level of .10, there is insufficient evidence to conclude that the mean test scores of the two classes are different. (Note that Minitab uses 8 degrees of freedom, rather than 7 degrees of freedom as is used in the textbook. The *P*-value will be slightly different.)

◀

▶ Exercise 19 (pg. 448) Tensile strength of steel bars

Open worksheet **Ex8_2-19.mtp**, which is found in the **ch08** MINITAB folder. The New method data is in C1 and the Old method data is in C2. Click on **Stat → Basic Statistics → 2-Sample t.** Select **Samples in different columns** and enter C1 for the **First** and C2 for the **Second** column. Select **Assume Equal Variances,** since the problem tells you to assume the population variances are equal. Click on **Options**, and then on the down arrow beside **Alternative** and select **not equal** since you want to test if the new treatment makes a difference in the tensile strength of steel bars. Click on **OK** twice and the results of the test should be displayed in the Session Window.

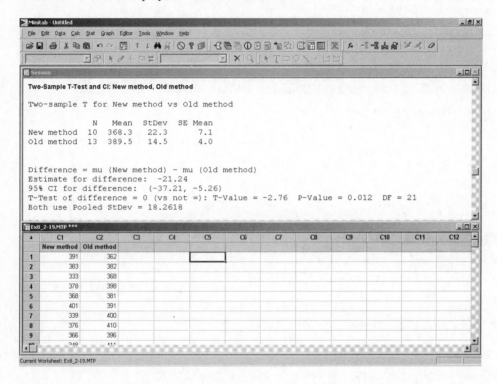

Since the *P*-value is 0.012 and is greater than the 0.01 significance level, there is insufficient evidence that the new method is different from the old method.

◀

> ▸ Exercise 21 (pg. 449) Comparing teaching methods

Open worksheet **Ex8_2-21.mtp**, which is found in the **ch08** MINITAB folder. The Old curriculum data is in C1 and the New curriculum data is in C2. Click on **Stat → Basic Statistics → 2-Sample t.** Select **Samples in different columns** and enter C1 for the **First** and C2 for the **Second** column. Select **Assume Equal Variances,** since the problem tells you to assume the population variances are equal. Click on **Options**, and then on the down arrow beside **Alternative** and select **less than** since you want to test if the students taught with the new curriculum had higher reading test scores than students taught with the old curriculum. Click on **OK** twice and the results of the test should be displayed in the Session Window.

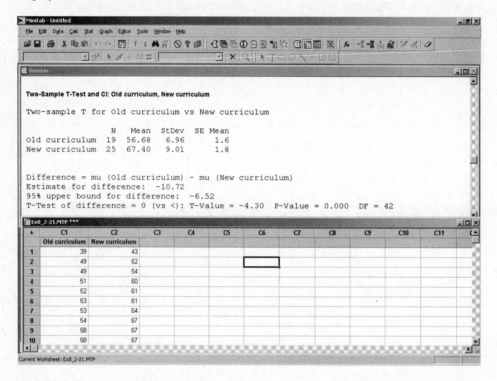

Since the *P*-value = 0.000, you would conclude that the new curriculum produces higher reading scores at all significance levels.

◀

Section 8.3

> ► Example 1 (pg. 453) Difference between means

Enter the data, found on page 453 of the textbook, into the MINITAB Data Window. Put the 'before' data in C1 and the 'after' data in C2. Click on **Stat → Basic Statistics → Paired t.** Enter C1 for the **First Sample** and C2 for the **Second Sample.**

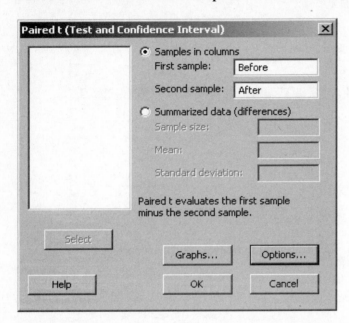

Click on **Options.** Enter 0 for **Test Mean** and select **less than** as the **Alternative.** Click on **OK** twice to display the results.

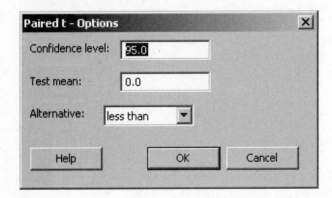

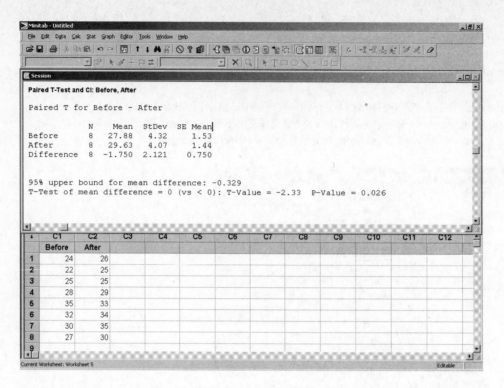

Since *p*=0.26, there is sufficient evidence at a significance level of 0.10 to conclude that the vertical heights are higher using the new shoes.

▶ Exercise 11 (pg. 457) Losing weight

Open worksheet **Ex8_3-11.mtp**, which is found in the **ch08** MINITAB folder. The 'before' weights are in C1 and the 'after' weights are in C2. Click on **Stat → Basic Statistics → Paired t.** Enter C1 for the **First Sample** and C2 for the **Second Sample.** Click on **Options.** Enter 0 for **Test Mean** and select **greater than** as the **Alternative** because, if the weights have decreased, then the difference (before - after) will be greater than 0. Click on **OK** twice to display the results.

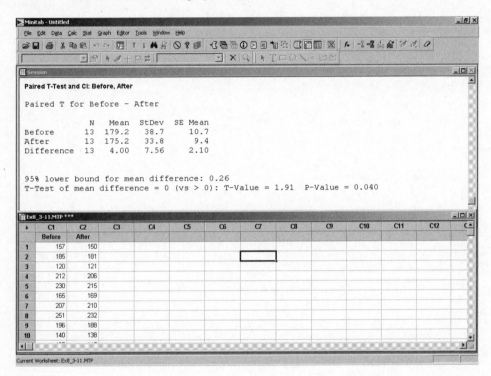

Since *p*=.04, you would conclude that the exercise program helps people lose weight at the 0.10 significance level.

◀

▶ Exercise 13 (pg. 457) Headaches

Open worksheet **Ex8_3-13.mtp**, which is found in the **ch08** MINITAB folder. The
'before' hours are in C1 and the 'after' hours are in C2. Click on **Stat → Basic Statistics**
→ Paired t. Enter C1 for the **First Sample** and C2 for the **Second Sample.** Click on
Options. Enter 0 for **Test Mean** and select **greater than** as the **Alternative** because, if
the weights have decreased, then the difference (before - after) will be greater than 0**.**
Click on **OK** twice to display the results.

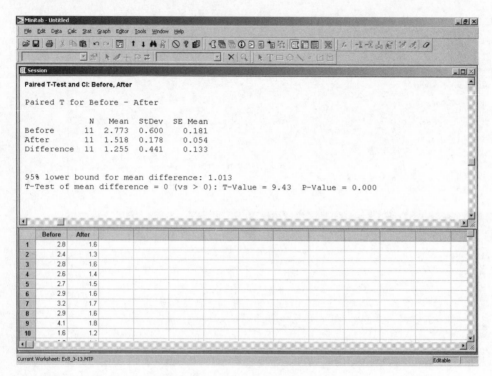

Since the *P*-value is so small, you should reject the null hypothesis. Thus, the
new therapy appears to reduce the duration of headaches.

Section 8.4

▸ Example 1 (pg. 463) Seat belt usage

In a study of 150 car drivers and 200 pickup truck drivers, 86% of the car drivers and 74% of pickup drivers were seat belts. This is a summary of the results of the study. To test if there is a difference in the proportion of seat belt users, click on **Stat → Basic Statistics → 2 Proportions.** Select **Summarized Data** and use the data for car drivers as the **First sample.** Enter 129 **Events** (150 x .86) and 150 **Trials**. Use the data for pickup drivers as the **Second sample.** Enter 148 **Events** (200 x .74) and 200 **Trials**.

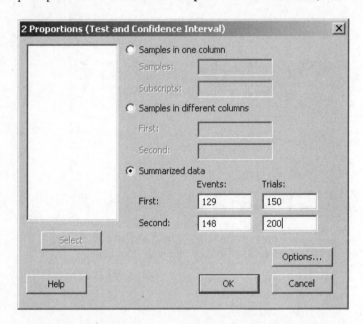

Click on **Options.** Enter 0 for **Test mean**, and select **not equal** as the **Alternative** since you want to test if there is a difference between the proportion of car and pickup drivers who wear seat belts. Next click on **Use pooled estimate of p for test.**

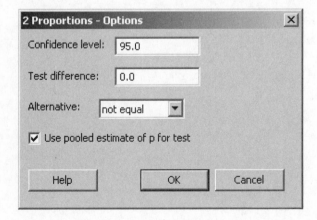

Click on **OK** twice to display the results in the Session Window.

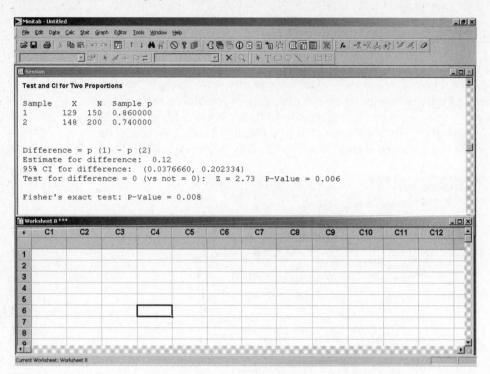

Since the *P*-value is smaller than α, you should Reject the null hypothesis and conclude that there is difference in the proportion of car and pickup drivers who wear seat belts.

▶ Exercise 13 (pg. 466) Migraines

To test if there is a difference in the proportion of adults who were pain-free after using a new medication is different from those using the placebo, click on **Stat → Basic Statistics → 2 Proportions.** Select **Summarized Data** and use the data for the new medication as the **First sample**. Enter 100 **Events** and 400 **Trials.** Use the placebo data as the **Second sample**. Enter 41 **Events** and 407 **Trials.** Click on **Options.** Enter 0 for **Test mean**, and select **not equal** as the **Alternative** since you want to test if there is a difference between the proportions. Next click on **Use pooled estimate of p for test.** Click on **OK** twice to display the results in the Session Window.

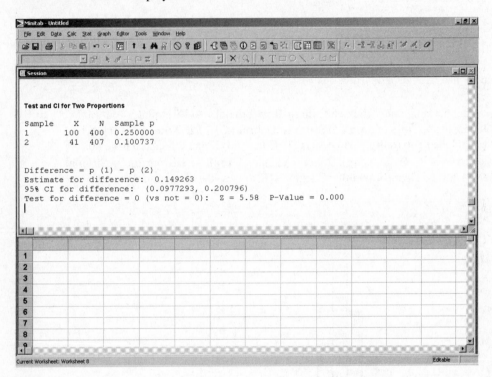

With such a small *P*-value, you should Reject the null hypothesis. Thus, the proportion of adults who are pain-free after using the new drug is different from those who used the placebo.

▸ Technology Lab (pg. 477) Tails over heads

1. Click on **Stat → Basic Statistics → 1 Proportion.** Select **Summarized Data** and
    enter 5772 **Events** and 11902 **Trials**. Click on **Options.** Enter .5 for **Test
    proportion**, and select **not equal** as the **Alternative** since you want to test if the
    probability is .5 or not. Next click on **Use test and interval based on normal
    distribution.** Click on **OK** twice to display the results in the Session Window.

```
Test and CI for One Proportion

Test of p = 0.5 vs p not = 0.5


Sample    X      N   Sample p    95% CI        Z-Value  P-Value
1       5772  11902  0.484961  (0.476, 0.494)   -3.28    0.001

Using the normal approximation.
```

3. To repeat the simulation, click on **Calc → Random data → Binomial.** You want
    to **Generate** 500 **rows of data** and **Store In Column** C1. The **Number of trials** is
    11902 and the **Probability of success** is .5. Click on **OK** and C1 should have 500
    rows of data in it. To draw the histogram, click on **Graph → Histogram → Simple.**
    Enter C1 for the **Graph variable**. Click on **OK** to view the histogram.

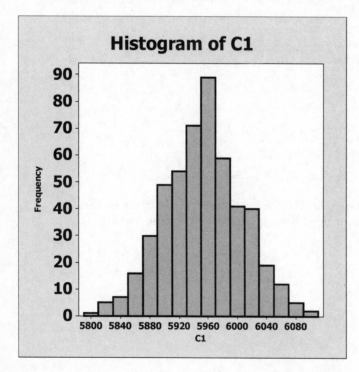

Correlation and Regression

Section 9.1

▶ Example 3 (pg. 486) Constructing a scatterplot

Open worksheet **OldFaithful,** which is found in the **ch09** MINITAB folder. The duration (in minutes) of several of Old Faithful's eruptions should be in C1, and the time (in minutes) until the next eruption should be in C2. Notice that Duration is the x-variable and Time is the y-variable. To plot the data, click on **Graph → Scatterplot → Simple.** On the input screen, enter C2 for the **Y variable** and C1 for the **X variable.**

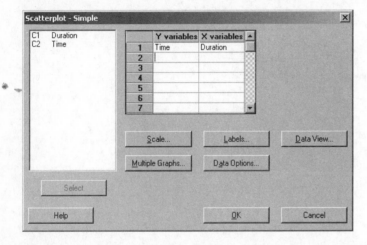

Next, click on **Labels** and enter a title for the plot. Click on OK to view the plot produced using Minitab default settings.

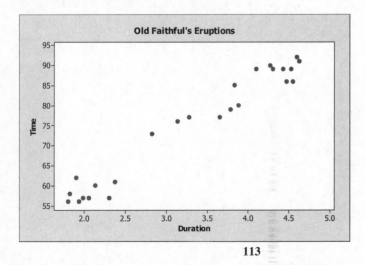

113

Click on **OK** twice to view the scatter plot. This plot would be better if both axes started at 0. To make this change, right-click on the X-axis scale and select "Edit X-scale". Enter 0 : 5 / .5 for **Position of ticks**. Click on **OK**. Next right-click on the Y-axis scale and select "Edit Y-scale". Enter 0 : 100 / 10 for **Position of ticks**. Click on **OK** to view the completed plot.

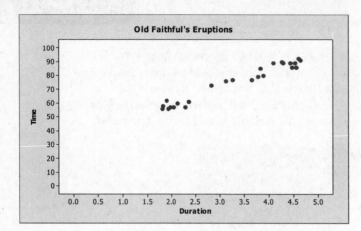

▶ Example 5 (pg. 489) Finding the correlation coefficient

Open worksheet **OldFaithful,** which is found in the **ch09** MINITAB folder. The
duration (in minutes) of several of Old Faithful's eruptions should be in C1, and the time
(in minutes) until the next eruption should be in C2. Notice that Duration is the x-
variable and Time is the y-variable. To find the correlation coefficient, click on **Stat →
Basics Statistics → Correlation.** On the input screen, select both C1 and C2 for
Variables, by double-clicking on each one.

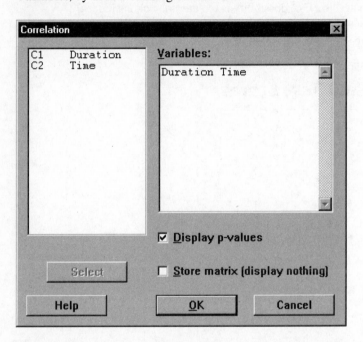

Click on **OK** and the correlation coefficient will be displayed in the Session Window.

Correlations: Duration, Time

```
Pearson correlation of Duration and Time = 0.979
P-Value = 0.000
```

▶ Exercise 23 (pg. 497) Plot of study hours vs. test scores

Open worksheet **Ex9_1-23,** which is found in the **ch09** MINITAB folder. The hours
spent studying should be in C1, and the test scores should be in C2. Notice that "Hours
study" is the x-variable and "Test Scores" is the y-variable. To plot the data, click on
Graph → Scatterplot → Simple. On the input screen, enter C2 for the **Y variable** and
C1 for the **X variable.** Next, click on **Labels** and enter a title for the plot. Click on OK
to view the plot produced using Minitab default settings. Edit the plot to set the tick mark
positions so that both axes begin at zero. One possibility is to enter the tick positions for
the **X variable** as 0:8/1 (0 to 8 in steps of 1) and for the **Y variable** as 0:100/10 (0 to 100
in steps of 10). Click on **OK** twice to view the scatter plot.

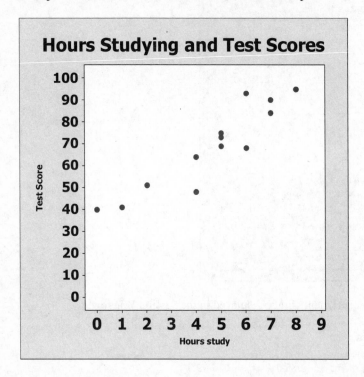

Now, to find the correlation coefficient, click on **Stat → Basics Statistics →
Correlation.** On the input screen, select both C1 and C2 for **Variables**, by double-
clicking on each one. Click on **OK** and the output should be in the Session Window.

```
Correlations: Hours studying, Test score
```

```
Pearson correlation of Hours studying and Test score = 0.923
P-Value = 0.000
```

The correlation between hours spent studying and test scores is 0.923 with a *p*-value of
0.000.

◀

Section 9.2

▸ Example 2 (pg. 503) Finding a regression equation

Open worksheet **OldFaithful,** which is found in the **ch09** MINITAB folder. The
duration (in minutes) of several of Old Faithful's eruptions should be in C1, and the time
(in minutes) until the next eruption should be in C2. Notice that Duration is the x-
variable and Time is the y-variable. To find the regression equation, click on **Stat →
Regression → Regression.** Enter C2 (Time) for the **Response** variable, and C1
(Duration) as the **Predictor.**

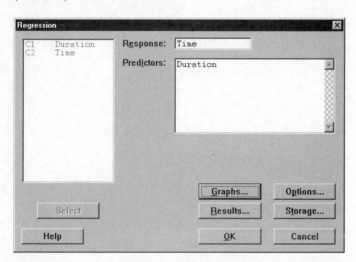

Click on **Results.** Select **Regression equation, table of coefficients, s, R-squared, and
basic analysis of variance.**

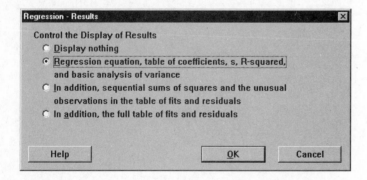

Click on **OK** twice to view the output in the Session Window.

Regression Analysis: Time versus Duration

The regression equation is
Time = 33.7 + 12.5 Duration

Predictor	Coef	SE Coef	T	P
Constant	33.683	1.894	17.79	0.000
Duration	12.4809	0.5464	22.84	0.000

S = 2.88153   R-Sq = 95.8%   R-Sq(adj) = 95.6%

Analysis of Variance

Source	DF	SS	MS	F	P
Regression	1	4331.7	4331.7	521.69	0.000
Residual Error	23	191.0	8.3		
Total	24	4522.6			

Notice that the regression equation is Time = 33.7 + 12.5 * Duration.

◀

▸ Exercise 20 (pg. 506)   Hours online vs. test scores

Open worksheet **Ex9_2-20,** which is found in the **ch09** MINITAB folder. Hours Online
is in C1 and Test is in C2. First plot the data. To plot the data, click on **Graph →
Scatterplot → Simple.** On the input screen, enter C2 for the **Y variable** and C1 for the
X variable. Next, click on **Labels** and enter a title for the plot. Click on OK to view the
plot produced using Minitab default settings. Edit the plot to set the tick mark positions
so that both axes begin at zero. Click on **OK** to view the scatter plot.

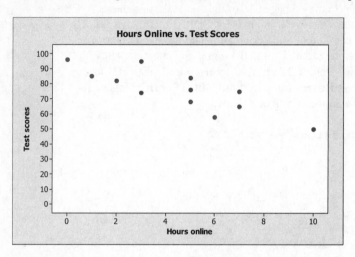

To find the regression equation, click on **Stat → Regression → Regression.** Enter C2
for the **Response** variable, and C1 as the **Predictor.** Click on **Results.** Select **Regression
equation, table of coefficients, s, R-squared, and basic analysis of variance.** Click on
OK twice to view the results in the Session Window.

Regression Analysis: Test scores versus Hours online
The regression equation is
Test scores = 94.0 - 4.07 Hours online

Predictor   Coef   SE Coef   T   P
Constant   93.970   4.524   20.77   0.000
Hours online   -4.0674   0.8600   -4.73   0.001

S = 8.11334   R-Sq = 69.1%   R-Sq(adj) = 66.0%

Analysis of Variance

Source   DF   SS   MS   F   P
Regression   1   1472.4   1472.4   22.37   0.001
Residual Error   10   658.3   65.8
Total   11   2130.7

Section 9.3

▸ Example 2 (pg. 516)   Finding the standard error and the
coefficient of determination

Enter the first two columns of data (found on page 516 of the textbook) into the
MINITAB Data Window. Enter the x's into C1 and name it GDP. Enter the y's into C2
and name it Emissions. Both the coefficient of determination and the standard error of
the estimate are part of the regression output. To find the regression equation, click on
Stat → Regression → Regression. Enter C2 for the **Response** variable, and C1 as the
Predictor. Click on **Results.** Select **Regression equation, table of coefficients, s, R-
squared, and basic analysis of variance.** Click on **OK** twice.

```
Regression Analysis: Emissions versus GDP

The regression equation is
Emissions = 102 + 196 GDP


Predictor     Coef   SE Coef      T      P
Constant    102.29     95.93   1.07  0.317
GDP         196.15     36.96   5.31  0.001


S = 138.255   R-Sq = 77.9%   R-Sq(adj) = 75.1%


Analysis of Variance

Source           DF       SS      MS      F      P
Regression        1   538235  538235  28.16  0.001
Residual Error    8   152916   19115
Total             9   691151
```

Notice that the standard error of the estimate is S = 138.255 and the coefficient of
determination is R-Sq = 77.9%.

▸ Example 3 (pg. 518)   Constructing a prediction interval

Enter the first two columns of data (found on page 516 of the textbook) into the
MINITAB Data Window. Enter the x's into C1 and name it GDP. Enter the y's into C2
and name it Emissions. Click on **Stat → Regression → Regression.** Enter C2 for the
Response variable, and C1 as the **Predictor.** To find both the point estimate and the
prediction interval, click on **Options. (Fit Intercept** is selected by default.) Next enter
the GDP. So enter 3.5 for **Prediction interval for new observations.** Enter 95 for the
Confidence level and select **Prediction limits.**

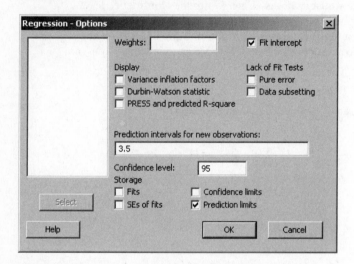

Click on **OK.** Click on **Results** and select **Regression equation, table of coefficients, s,
R-squared, and basic analysis of variance.** Click on **OK** twice and view the output in
the Session Window. Below are the results of the prediction interval only. The
regression output is the same as was produced in Example 2.

Regression Analysis: Emissions versus GDP

Predicted Values for New Observations

```
New
Obs    Fit   SE Fit      95% CI              95% PI
 1    788.8   62.0   (645.8, 931.8)   (439.4, 1138.2)
```

Values of Predictors for New Observations

```
New
Obs   GDP
 1   3.50
```

Notice that the predicted value, 788.8, is listed below **Fit** and the prediction interval,
(439.4, 1138.2), is listed below **95% PI.**

◀

▶ Exercises 13 (pg. 520) and 21 (pg. 521) Retail space vs. sales

Open worksheet **Ex9_3-13,** which is found in the **ch09** MINITAB folder. Square footage
is in C1 and Sales is in C2. For these two problems, you need to find the coefficient of
determination, the standard error of the estimate, and a 90% prediction interval when
square footage is 5.75 billion. This can be accomplished all at once in MINITAB. Click
on **Stat → Regression → Regression.** Enter C2 for the **Response** variable, and C1 as
the **Predictor.** Now, to find both the point estimate and the prediction interval, click on
Options. Select **Fit Intercept** by clicking on it. Next enter the square footage. So enter
5.75 for **Prediction interval for new observations.** Enter 90 for the **Confidence level**
and select **Prediction limits.** Click on **OK.** Click on **Results.** Since you would like to
see the other regression output, then select **Regression equation, table of coefficients, s,
R-squared, and basic analysis of variance.** Click on **OK** twice and view the output in
the Session Window.

```
Regression Analysis:
Sales (in billions of dollars versus Square footage)

The regression equation is
Sales (in billions of dollars) = - 1882 + 549 Square footage

Predictor           Coef   SE Coef        T       P
Constant         -1881.7     143.8   -13.09   0.000
Square footage    549.45     25.79    21.30   0.000


S = 30.5759    R-Sq = 98.1%   R-Sq(adj) = 97.8%

Analysis of Variance

Source          DF       SS       MS        F       P
Regression       1   424297   424297   453.85   0.000
Residual Error   9     8414      935
Total           10   432711

Predicted Values for New Observations
New
Obs     Fit   SE Fit          90% CI                90% PI
  1  1277.63   10.40   (1258.58, 1296.69)   (1218.43, 1336.83)

Values of Predictors for New Observations

New    Square
Obs    footage
  1      5.75
```

From the output, you can see that the coefficient of determination is 0.981, the standard
error of the estimate is 30.5759, and a 90% prediction interval when square footage is
5.75 billion is (1218.43, 1336.83).

◀

Section 9.4

▸ Example 1 (pg. 524) Finding a multiple regression equation

Open worksheet **Salary,** which is found in the **ch09** MINITAB folder.
Salary should be in C1, Employment in C2, Experience in C3, and Education in C4.
Click on **Stat** → **Regression** → **Regression.** Enter C1 for the **Response** variable, and
enter C2, C3, and C4 as the **Predictors.** Click on **Results.** Since you would like to see
the other regression output, then select **Regression equation, table of coefficients, s, R-
squared, and basic analysis of variance.** Click on **OK** twice and view the output in the
Session Window.

```
Regression Analysis: Salary versus Employment, Experience,
and Education


The regression equation is
Salary = 49764 + 364 Employment (yrs) + 228 Experience (yrs)
         + 267 Education (yrs)


Predictor          Coef   SE Coef      T      P
Constant          49764      1981  25.12  0.000
Employment (yrs)  364.41     48.32   7.54  0.002
Experience (yrs)  227.6      123.8   1.84  0.140
Education (yrs)   266.9      147.4   1.81  0.144


S = 659.490    R-Sq = 94.4%   R-Sq(adj) = 90.2%


Analysis of Variance

Source          DF        SS       MS      F      P
Regression       3  29231989  9743996  22.40  0.006
Residual Error   4   1739708   434927
Total            7  30971697
```

The regression equation is listed at the beginning of the output. Notice that the regression
equation uses the values listed below **Coef.** These values are the coefficients of the
multiple regression equation.

◀

▶ Exercise 5 (pg. 528) Finding a multiple regression equation

Open worksheet **Ex9_4-5,** which is found in the **ch09** MINITAB folder.
Sales should be in C1, Square footage in C2, and Number of shopping centers in C3.
Click on **Stat → Regression → Regression.** Enter C1 for the **Response** variable, and
enter C2 and C3 as the **Predictors.** Click on **Results.** Since you would like to see the
other regression output, then select **Regression equation, table of coefficients, s, R-squared, and basic analysis of variance.** Click on **OK** twice and view the output in the
Session Window.

Regression Analysis: Sales versus Square footage, Shopping centers

```
The regression equation is
Sales = - 2518 + 127 Square footage + 66.4 Shopping centers


Predictor              Coef  SE Coef       T      P
Constant            -2518.4    435.0   -5.79  0.000
Square footage        126.8    275.8    0.46  0.658
Shopping centers      66.36    43.13    1.54  0.162


S = 28.4890   R-Sq = 98.5%   R-Sq(adj) = 98.1%


Analysis of Variance

Source          DF       SS      MS       F      P
Regression       2   426218  213109  262.57  0.000
Residual Error   8     6493     812
Total           10   432711
```

The regression equation is listed at the beginning of the output:

Sales = - 2518 + 127 * Square footage + 66.4 * Shopping centers.

> ▶ Technology Lab (pg. 537) Sugar, fat, and carbohydrates

Open worksheet **Tech9,** which is found in the **ch09** MINITAB folder.
Cereal Name is in C1, Calories is in C2, Sugar is in C3, Fat is in C4, and Carbohydrates
is in C5.

1. Click on **Graph → Scatterplot → Simple.** On the input screen, enter C2 for the **X
    variable** and C3 for the **Y variable.** Click on **Labels** and enter a title. Click on **OK.**
    Repeat these steps for parts b - f, changing the variables as directed in each part.

3. Click on **Stat → Basics Statistics → Correlation.** On the input screen, select both
    C2 and C3 for **Variables** by double-clicking on each one. Click on **OK** and the
    output should be in the Session Window. Repeat these steps for each pair of
    variables listed in question 1, parts b - f.

4. (Do both 4 & 5 at one time here). Click on **Stat → Regression → Regression.**
    Enter C3 for the **Response** variable, and enter C2 as the **Predictor.** Next, to predict
    the sugar content of 1 cup of cereal with a calorie content of 120 kcal, click on
    Options. Select **Fit Intercept** by clicking on it. Next enter the calorie content. So
    enter 120 for **Prediction interval for new observations.** Enter 95 for the
    Confidence level and select **Prediction limits.** Click on **OK.** Next, click on
    Results. Since you would like to see the other regression output, select **Regression
    equation, table of coefficients, s, R-squared, and basic analysis of variance.**
    Click on **OK** twice and view the output in the Session Window. Repeat these steps
    using C4 for the **Response** variable.

5. (Do both 6 & 7 at one time here). Click on **Stat → Regression → Regression.**
    Enter C2 for the **Response** variable, and enter C3, C4, and C5 as the **Predictors.**
    Next, click on **Results.** Since you would like to see the other regression output, then
    select **Regression equation, table of coefficients, s, R-squared, and basic analysis
    of variance.** Click on **OK** twice and view the output in the Session Window. For
    part b, repeat these steps using C3 and C5 for the **Predictors.** Next, to predict the
    calorie content of 1 cup of cereal with 10g of sugar and 25g of carbohydrates, click
    on **Options.** Select **Fit Intercept** by clicking on it. Next enter the sugar and
    carbohydrate contents. To enter both of these, type in a 10 (for the sugar), leave a
    space, and then type in a 25 (for the carbohydrate) for **Prediction interval for new
    observations.** Click on **OK.** Next, click on **Results.** Since you would like to see
    the other regression output, then select **Regression equation, table of coefficients, s,
    R-squared, and basic analysis of variance.** Click on **OK** twice and view the
    output in the Session Window.

◀

Section 10.1

▶ Example 3 (pg. 544) Chi-square goodness-of-fit test

Enter the data into the MINITAB Data Window. Enter the Colors into C1 and name it Response. Enter the observed frequencies into C2 and name it Observed. In this example, it is assumed to be a uniform distribution, so each color should be about 1/6th (0.167) of the bag.

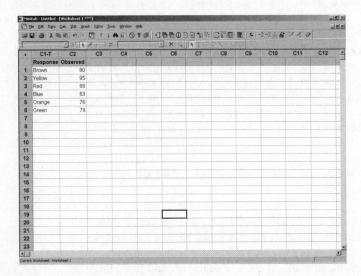

To perform the Goodness of Fit test, click on **Stat → Tables → Chi-square Goodness of Fit Test**. Select **Observed counts** and enter C2 (Observed). Click on the input box beside **Category names** and enter C1 (Response). Select **Equal proportions**.

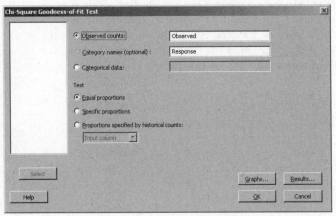

Click on **OK** and the Chi-Square test statistic will be displayed in the Session Window.

Chi-Square Goodness-of-Fit Test for Observed Counts in Variable: Observed

Using category names in Response

Category	Observed	Test Proportion	Expected	Contribution to Chi-Sq
Brown	80	0.166667	83.3333	0.13333
Yellow	95	0.166667	83.3333	1.63333
Red	88	0.166667	83.3333	0.26133
Blue	83	0.166667	83.3333	0.00133
Orange	76	0.166667	83.3333	0.64533
Green	78	0.166667	83.3333	0.34133

N	DF	Chi-Sq	P-Value
500	5	3.016	0.698

Since the test statistic is 3.016 and the P-value is 0.698, which is larger than $\alpha = .05$, you would fail reject the null hypothesis.

◀

► Exercise 13 (pg. 548)   College education

Enter the data into the MINITAB Data Window. Enter the survey Responses into C1 and
name it Response. Enter the observed frequencies into C2 and name it Observed. In this
example, you want to compare the responses from the teenagers to the distribution of
parental responses, so enter the distribution of parental responses shown in the pie chart
into C3 and name it Distribution. Be sure to enter the percents shown in the pie chart as
proportions (i.e. 55% = 0.55).

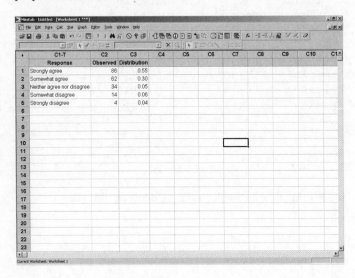

To perform the goodness-of-fit test, click on **Stat → Tables → Chi-square Goodness of
Fit Test**. Select **Observed counts** and enter C2 (Observed). Click on the input box
beside **Category names** and enter C1 (Response).   Select **Specific proportions** and
enter C3 (Distribution).

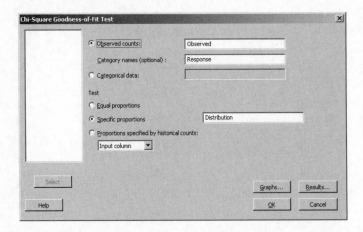

Click on **OK** and the chi-square test statistic will be displayed in the Session Window.

```
Chi-Square Goodness-of-Fit Test for Observed Counts in Variable: Observed

Using category names in Response

                                           Test                     Contribution
Category                       Observed  Proportion  Expected        to Chi-Sq
Strongly agree                       86       0.55       110          5.2364
Somewhat agree                       62       0.30        60          0.0667
Neither agree nor disagree           34       0.05        10         57.6000
Somewhat disagree                    14       0.06        12          0.3333
Strongly disagree                     4       0.04         8          2.0000


  N  DF   Chi-Sq  P-Value
200   4  65.2364    0.000
```

In this example, the test statistic is 65.2364 and the P-value=0.000. Since this P-value is so small, you would reject the null hypothesis at any α-level.

◀

Section 10.2

▶ Example 3 (pg. 556) Chi-square independence test

Enter the data into the MINITAB Data Window. First label the columns: use Gender for C1, "0-1" for C2, ... "6-7" for C5. Now enter the data into the appropriate columns. Do not enter any totals.

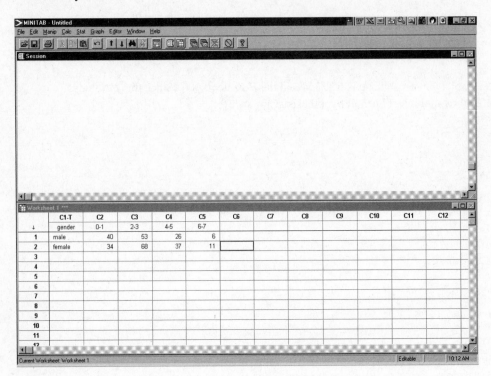

To perform the chi-square independence test, click on **Stat → Tables → Chi-square Test (Two way table in worksheet).** On the input screen, select C2 - C5 for the **Columns containing the table.** Click on **OK** and the test results will be displayed in the Session Window.

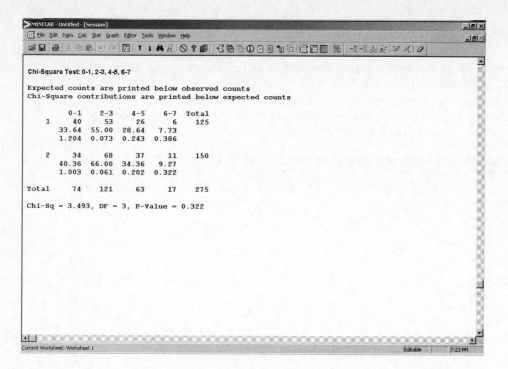

```
MINITAB - Untitled - [Session]
File  Edit  Data  Calc  Stat  Graph  Editor  Tools  Window  Help

Chi-Square Test: 0-1, 2-3, 4-5, 6-7

Expected counts are printed below observed counts
Chi-Square contributions are printed below expected counts

           0-1      2-3      4-5      6-7   Total
     1      40       53       26        6     125
         33.64    55.00    28.64     7.73
          1.204    0.073    0.243    0.386

     2      34       68       37       11     150
         40.36    66.00    34.36     9.27
          1.003    0.061    0.202    0.322

Total      74      121       63       17     275

Chi-Sq = 3.493, DF = 3, P-Value = 0.322
```

Notice that the test statistic is Chi-Sq = 3.493 and the *P*-value = .322. Since this *P*-value is larger than α = .05, you should not reject the null hypothesis. Thus, there is not enough evidence to conclude that the number of days per week spent exercising is related to gender.

▶ Exercise 17 (pg. 559) Should the drug be used as treatment?

Enter the data into the MINITAB Data Window. Enter Result data into C1, Drug data
into C2 and Placebo data into C3. To perform the chi-square independence test, click on
Stat → Tables → Chi-square Test (Two way table in worksheet). On the input
screen, select C2 - C3 for the **Columns containing the table.** Click on **OK** and the test
results will be displayed in the Session Window.

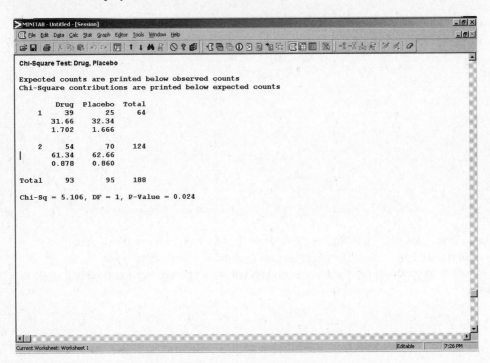

Notice that the test statistic is Chi-Sq = 5.106 and the *P*-value = .024. Since this *P*-value
is smaller than $\alpha = .10$, you should reject the null hypothesis. Thus, there is evidence to
conclude that the drug should be used as part of the treatment.

◀

Section 10.3

▶ Example 3 (pg. 569) Performing a two-sample *F*-test

A restaurant manager designed a system to decrease the variance in customer wait times. To perform the two-sample *F*-test, click on **Stat → Basic Statistics → 2 Variances.** Click on **Summarized data**, since this example has only sample sizes and variances. Beside **First**, enter the Sample size and Variance for the old system and beside **Second**, enter the data for the new system. Click on **OK** to see the results of the *F*-test in both the Graph Window and in the Session Window. Below are the results from the Session Window.

```
Test for Equal Variances

95% Bonferroni confidence intervals for standard deviations

Sample   N    Lower   StDev    Upper
     1  10  13.0825     20   40.2731
     2  21  11.8017     16   24.4620


F-Test (Normal Distribution)
Test statistic = 1.56, p-value = 0.388
```

The *F*-test results show that the test statistic = 1.56 and the *P*-value = .388. Since this is a large *P*-value, you would fail to reject the null hypothesis. Thus, the two variances are approximately equal.

◀

▶ Exercise 21 (pg. 572)   Home theater prices

Enter the data into the MINITAB Data Window. Enter the Company A prices into C1 and Company B prices into C2. In this example, you want to compare the variances of the prices between the two companies.

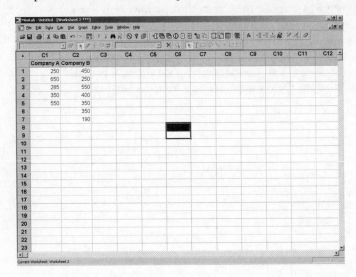

Click on **Samples in different columns** and enter C1 for **First** and C2 for **Second**. Click on **OK** to see the results of the *F*-test in both the Graph Window and in the Session Window. Below are the results from the Session Window.

Test for Equal Variances: Company A, Company B

```
95% Bonferroni confidence intervals for standard deviations

              N    Lower    StDev    Upper
Company A   5   97.6857  174.485  603.639
Company B   7   73.1583  120.376  302.808


F-Test (Normal Distribution)
Test statistic = 2.10, p-value = 0.398
```

The *F*-test results show that the test statistic = 2.10 and the *P*-value = .398. Since this is a large *P*-value, you would fail to reject the null hypothesis. Thus, the two variances are approximately equal.

◀

Section 10.4

> ▶ Example 2 (pg. 579) ANOVA tests

A researcher believes the earnings of top-paid actors, athletes, and musicians are the same. Enter the data found on page 579 of the textbook into C1, C2, and C3.

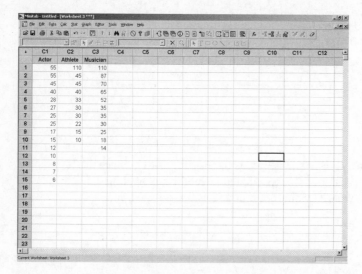

To perform a one-way analysis of variance, click on **Stat → ANOVA → One Way (Unstacked).** Select all three columns for **Responses (in separate columns)** and click on **OK.** The results of the test will be in the Session Window.

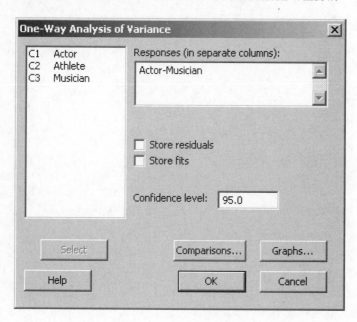

```
One-way ANOVA: Actor, Athlete, Musician

Source  DF     SS    MS     F      P
Factor   2   3766  1883  3.05  0.061
Error   33  20364   617
Total   35  24130

S = 24.84    R-Sq = 15.61%   R-Sq(adj) = 10.49%

                                Individual 95% CIs For Mean Based on
                                Pooled StDev
Level       N   Mean  StDev   --+---------+---------+---------+-------
Actor      15  25.00  16.84   (--------*-------)
Athlete    10  38.00  27.90         (---------*----------)
Musician   11  49.18  30.64                 (---------*---------)
                                --+---------+---------+---------+-------
                                  15        30        45        60

Pooled StDev = 24.84
```

Notice that the test statistic is listed ($F = 3.05$), as well as the *P*-value (.061). Since the *P*-value = .061, and this is smaller than $\alpha = .10$, you should reject the null hypothesis. Thus, there is a difference in the average earnings of top-paid actors, athletes, and musicians.

▶ Exercise 5 (pg. 581)   Costs per month of different toothpastes

Open worksheet **Ex10_4-5** found in the MINITAB folder **ch10.** The data for the cost per ounces for three qualities of stain removal should be in C1, C2, and C3. To perform a one-way analysis of variance, click on **Stat → ANOVA → One Way (Unstacked).** Select all three columns and click on **OK.** The results of the test will be in the Session Window.

```
One-way ANOVA: Very Good, Good, Fair

Source  DF     SS     MS     F      P
Factor   2  0.518  0.259  1.02  0.376
Error   26  6.629  0.255
Total   28  7.148

S = 0.5049   R-Sq = 7.25%   R-Sq(adj) = 0.12%


                            Individual 95% CIs For Mean Based on
                            Pooled StDev
Level       N    Mean   StDev  ---+---------+---------+---------+------
Very Good   12  0.5358  0.3222    (-----------*-----------)
Good        12  0.8267  0.6657                (-----------*-----------)
Fair         5  0.6300  0.3913  (-----------------*------------------)
                               ---+---------+---------+---------+------
                               0.25      0.50      0.75      1.00

Pooled StDev = 0.5049
```

Since the *P*-value = .376 and is larger than α, you should not reject the null hypothesis. Thus, there is not enough evidence that the average cost per month is the different for the three types of toothpaste.

◀

▶ Exercise 11 (pg. 583) Days spent in a hospital

Open worksheet **Ex10_4-11** found in the MINITAB folder **ch10.** The data for the four
different regions of the United States should be in C1 - C4. To perform a one-way
analysis of variance, click on **Stat → ANOVA → One-Way (Unstacked).** Select all
four columns and click on **OK.** The results of the test will be in the Session Window.

One-way ANOVA: Northeast, Midwest, South, West

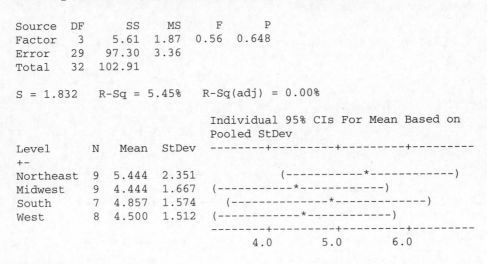

```
Source  DF      SS     MS      F      P
Factor   3    5.61   1.87   0.56   0.648
Error   29   97.30   3.36
Total   32  102.91

S = 1.832   R-Sq = 5.45%   R-Sq(adj) = 0.00%

                          Individual 95% CIs For Mean Based on
                          Pooled StDev
Level       N   Mean  StDev  --------+---------+---------+---------
+-
Northeast   9  5.444  2.351                (-----------*------------)
Midwest     9  4.444  1.667  (-----------*------------)
South       7  4.857  1.574     (--------------*--------------)
West        8  4.500  1.512  (------------*------------)
                             --------+---------+---------+---------
                               4.0       5.0       6.0


Pooled StDev = 1.832
```

Since the *P*-value = .648 and is larger than α, you should NOT reject the null hypothesis.
Thus, the average number of days spent in a hospital is the same for all four regions of
the United States.

◀

▸ Technology Lab (pg. 595)   Teacher salaries

Open worksheet **Tech10_a** found in the MINITAB folder **ch10.** The data for the teacher salaries in three states should be in C1 - C3.

1.   Since the three states represent different populations and the three samples were randomly chosen, the samples are independent.

2.   MINITAB has a test for normality. Click on **Stat → Basic Statistics → Normality Test.** Select C1 for the **Variable** and **Kolmogorov-Smirnov** for the **Test of Normality.** Click on **OK** and a normal plot will be displayed.

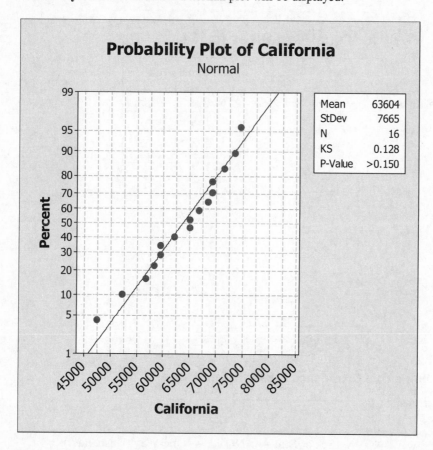

Notice the P-value is listed with results. Since this P-value is larger than .10 (our α), then you can assume that the data is approximately normal. Repeat this test for the other two columns of data.

3.   To test if the 3 samples have equal variances, MINITAB requires that the data be stacked into one column with a second column identifying which sample each data value came from. To do this, click on **Data → Stack → Columns.** Select all three

columns to be stacked on top of each other. **Column of current worksheet** should be C4 and **Store subscripts in** C5. The subscripts will be numbers 1 or 2 or 3 to indicate which column the data value came from, or if you select **Use variable names in subscript column** then the column names will be your subscripts. Next perform the Test for Equal Variances. To do this, click on **Stat → ANOVA → Test for Equal Variances.** On the input screen, C4 is the **Response** variable and C5 is the **Factor.** Enter an appropriate **Title** and click on **OK.** This test produces quite a lot of output, however you are only interested in the results of the *F*-test. You can see the results in both the Session Window and the Graph Window. Below are the results from the Graph Window. Bartlett's Test assumes normality of the data. Since the *P*-values are large, you can assume that the variances are approximately equal.

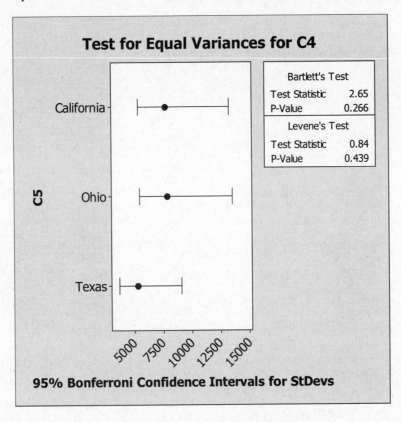

4. To do a one-way ANOVA, click on **Stat → ANOVA → One-Way (Unstacked).** Select all three columns and click on **OK.** The results of the test will be in the Session Window. Look at the P-value. If it is smaller than α, you should reject the null hypothesis.

5. Repeat Exercises 1 - 4 using worksheet **Tech10_b** found in the MINITAB folder **ch10.**

Nonparametric Tests

C H A P T E R

11

Section 11.1

▶ Example 3 (pg. 603) Using the paired-sample sign
test

Open worksheet **Rehabilitation** which is found in the **ch11** MINITAB folder. The
'Before' data is in C1 and the 'After' data is in C2. Since we are interested in the
difference between C1 and C2, we will create a new column that is C1 - C2. Click on
Calc → Calculator. Store result in variable C3 and calculate the **Expression** C1 - C2.
Click on **OK** and C3 should contain the differences. Now perform a one-sample Sign
test on C3. Click on **Stat → Nonparametrics → 1-sample Sign.** Select C3 as the
Variable. Since you would like to test if the number of repeat offenders has decreased
after the special course, you would expect that the differences in C3 would be greater
than 0. Thus, enter 0 for **Test median** and select **greater than** for the **Alternative.**

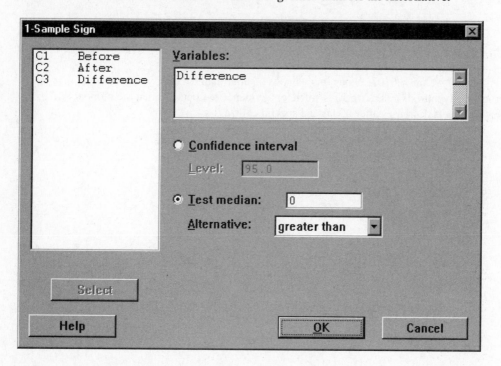

Click on **OK** and the results will be in the Session Window.

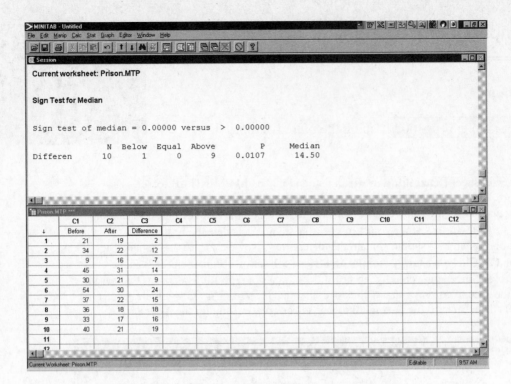

Notice that the *p*-value = .0107. Since this value is smaller than α=.025, you would reject the null hypothesis. Thus, there is sufficient evidence to conclude that the number of repeat offenders decreases after taking the special course.

▶ Exercise 7 (pg. 604)   Credit card charges

Enter the data into C1 in the MINITAB Data Window. (Do not type in the $ sign.) To perform the Sign Test, click on **Stat → Nonparametrics → 1-sample Sign.** Select C1 as the **Variable.** Since you would like to test if the median amount of new credit card charges was more than $300, enter 300 for **Test median** and select **greater than** for the **Alternative.**

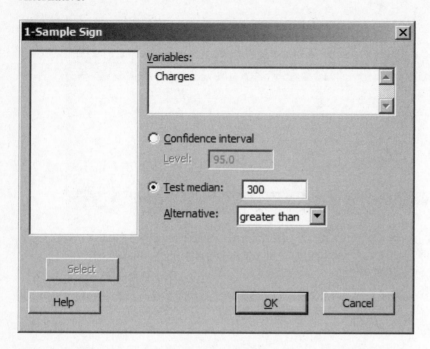

Click on **OK** and the results will be displayed in the Session Window.

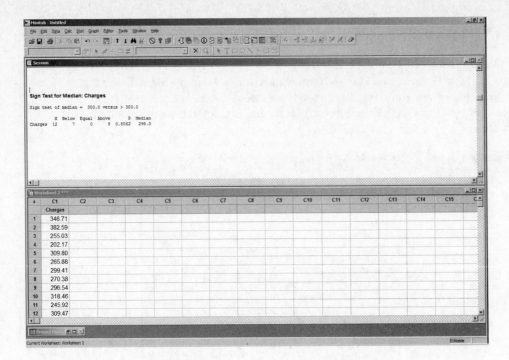

Notice that the *p*-value = .8062. Since this is such a large *p*-value, you would fail to reject the null hypothesis. Thus, the accountant can **not** conclude that the median amount of new charges was more than $300.

▸ Exercise 20 (pg. 606) Does therapy decrease pain intensity?

Open worksheet **Ex11_1-20** which is found in the **ch11** MINITAB folder. Pain intensity before taking anti-inflammatory drugs is in C1 of the MINITAB Data Window and Pain intensity after taking the drug is in C2. Calculate the differences. Click on **Calc →**
Calculator. Next **Store result in variable** C3 and calculate the **Expression** C1-C2.
Click on **OK** and C3 should contain the differences. Now perform a one-sample Sign test on C3. Click on **Stat → Nonparametrics → 1-sample Sign.** Select C3 as the **Variable.** Since you would like to test if the intensity scores have decreased after the anti-inflammatory drugs, you would expect that the differences in C3 would be greater than 0. Thus, enter 0 for **Test median** and select **greater than** for the **Alternative.**
Click on **OK.**

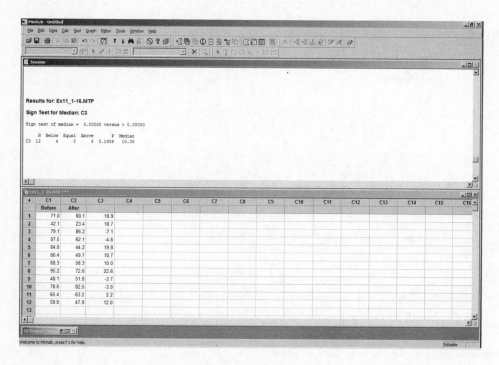

Notice that the p-value = .1938. Since this value is larger than α=.05, you would fail to reject the null hypothesis. Thus, there is not enough evidence to conclude that the pain intensity decreases with these drugs.

◀

Section 11.2

▶ Example 1 (pg. 610) Performing a Wilcoxon signed-rank test

Enter the 'Old Design' data in C1 and 'New design' data in C2. To calculate the differences, click **Calc → Calculator. Store the result in variable** C3 and calculate the **Expression** C1-C2. Click on **OK** and C3 should contain the differences. To perform the Wilcoxon Signed Rank Test, click on **Stat → Nonparametrics → 1-sample Wilcoxon.** You should use C3 for the **Variable.** Since you are using the differences in this example, you want to compare the median difference to 0. So, enter 0 for **Test Median** and choose **greater than** as the **Alternative** since the claim is that the new design lowers scores. (If Old – New > 0 then the New scores are lower than the old scores.)

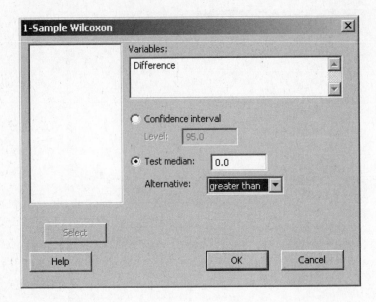

Click on **OK** to view the results of the test in the Session Window.

```
Wilcoxon Signed Rank Test: Difference
Test of median = 0.000000 versus median > 0.000000

                  N for  Wilcoxon              Estimated
            N     Test   Statistic      P      Median
Difference  10     9       38.5      0.033     2.500
```

The MINITAB output tells you that the Wilcoxon Statistic is 38.5 and the *p*-value is .033. Although the textbook tells you to select the smaller of the absolute value of the two sums of the ranks, MINITAB simply uses the sum of the positive ranks. This makes no difference in interpreting the results. The important thing to notice is that the *p*-value = .033. Since this is smaller than α = .05, you would reject the null hypothesis. Thus, there is sufficient evidence to say that golfers can lower their scores with the newly designed clubs.

◀

▶ Example 2 (pg. 613) Performing a Wilcoxon rank sum test

Open worksheet **Earnings** which is found in the **ch11** MINITAB folder. Male earnings
are in C1 and Female earnings are in C2. In MINITAB, the Wilcoxon Rank Sum Test is
called the Mann-Whitney test. Click on **Stat → Nonparametrics → Mann-Whitney.**
Enter C1 for the **First Sample,** C2 for the **Second Sample**, and enter 90.0 for the
Confidence level and select n**ot equal** as the **Alternative** since you want to see if there is
a difference between the earnings.

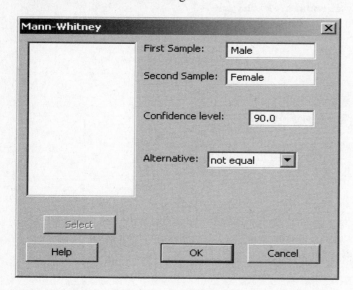

Click on **OK** and the results will be in the Session Window.

Mann-Whitney Test and CI: Male, Female

```
          N   Median
Male     10    96.50
Female   12    90.50


Point estimate for ETA1-ETA2 is 7.00
90.7 Percent CI for ETA1-ETA2 is (-1.00,14.00)
W = 138.0
Test of ETA1 = ETA2 vs ETA1 not = ETA2 is significant at 0.1379
The test is significant at 0.1375 (adjusted for ties)
```

Look at the results carefully. The P-value is .1379. The rank sum for male earnings is
also listed, W = 138. In this example, since the *p*-value is larger than α=.10, you would
fail to reject the null hypothesis. Thus, there is no difference between male and female
earnings.

◀

▶ Exercise 7 (pg. 616) Teacher salaries

Open worksheet **Ex11_2-7** which is found in the **Chapter 11** MINITAB folder. Click on
Stat → Nonparametrics → Mann-Whitney. Enter C1 for the **First Sample,** C2 for the
Second Sample, and select n**ot equal** as the **Alternative** since you want to see if there is
a difference between the salaries in Wisconsin and Michigan. Click on **OK**. The results
should be in the Session Window.

Mann-Whitney Test and CI: Wisconsin, Michigan

```
            N   Median
Wisconsin   11  51.000
Michigan    12  57.000


Point estimate for ETA1-ETA2 is -4.500
95.5 Percent CI for ETA1-ETA2 is (-10.002,0.002)
W = 100.5
Test of ETA1 = ETA2 vs ETA1 not = ETA2 is significant at 0.0564
The test is significant at 0.0555 (adjusted for ties)
```

Since the p-value = .0564 and is larger than α=.05, you would fail to reject the null
hypothesis. There is not enough evidence to conclude there is a difference in teacher
salaries in Wisconsin and Michigan.

◀

Section 11.3

▶ Example 1 (pg. 621) Performing a Kruskal-Wallis test

Open worksheet **Crimes** which is found in the **ch11** MINITAB folder. To perform a Kruskal-Wallis test, MINITAB requires that the data be stacked into one column with a second column identifying which sample each data value came from. To do this, click on **Data → Stack → Columns.** Select all three columns to be stacked on top of each other. Select **Column of current worksheet** and enter C4 and **Store subscripts in** C5. The subscripts will be numbers 1, 2, or 3 to indicate which column the data value came from. Be sure that **Use variable names in subscript column** is NOT selected.

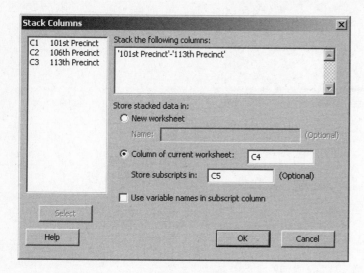

Click on **OK.** Name C4 Crimes and C5 Precinct. Notice that in C5, 1 represents the 101st, 2 represents the 106th, and 3 represents the 113th precinct.

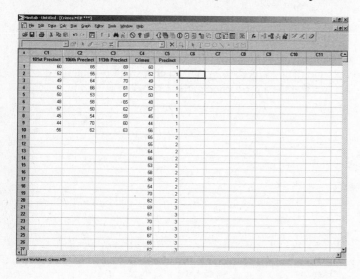

Now, to do the Kruskal-Wallis test, click on **Stat → Nonparametrics → Kruskal-Wallis.** The **Response** variable is Crimes (C4) and the **Factor** is Precinct (C5).

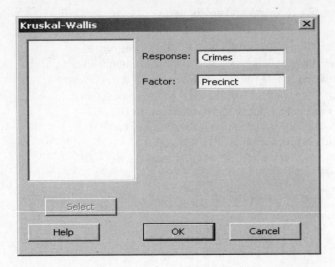

Click on **OK** and the results will be displayed in the Session Window.

Results for: Crimes.MTP

Kruskal-Wallis Test: Crimes versus Precinct

Kruskal-Wallis Test on Crimes

```
Precinct   N   Median  Ave Rank       Z
1         10   51.00        7.7   -3.43
2         10   60.00       17.7    0.97
3         10   62.50       21.1    2.46
Overall   30               15.5

H = 12.52  DF = 2  P = 0.002
H = 12.54  DF = 2  P = 0.002  (adjusted for ties)
```

Notice that the test statistic is H=12.52 and the *p*-value=.002. With such a small *p*-value, you should reject the null hypothesis. So, there is a difference in the number of crimes in the three precincts.

▸ Exercise 3 (pg. 623) Are the insurance premiums different?

Open worksheet **Ex11_3-3** which is found in the **ch11** MINITAB folder. The data
should be in C1, C2, and C3. To perform a Kruskal-Wallis test, MINITAB requires that
the data be stacked into one column with a second column identifying which sample each
data value came from. To do this, click on **Data → Stack → Columns.** Select all three
columns to be stacked on top of each other. Select **Column of current worksheet** and
enter C4 and **Store subscripts in** C5. The subscripts will be numbers 1, 2, or 3 to
indicate which column the data value came from. Be sure that **Use variable names in
subscript column** is NOT selected. Click on **OK.** Name C4 Premiums and C5 State.
Notice that in C5, 1 represents Connecticut, 2 represents Massachusetts, and 3 represents
Virginia. Now, to do the Kruskal-Wallis test, click on **Stat → Nonparametrics →
Kruskal-Wallis.** The **Response** variable is Premiums (C4) and the **Factor** is State (C5).
Click on **OK** and the results will be displayed in the Session Window.

Kruskal-Wallis Test: Premiums versus State

```
Kruskal-Wallis Test on Premiums

State     N   Median  Ave Rank      Z
1         7    930.0      12.6   0.82
2         7   1025.0      15.1   2.16
3         7    625.0       5.3  -2.98
Overall  21              11.0

H = 9.51   DF = 2   P = 0.009
```

Notice that the test statistic is H=9.51 and the p-value=.009. With such a small p-value,
you should reject the null hypothesis. So, there is a difference in the premiums of the
three states.

◂

Section 11.4

> ▶ **Example 1 (pg. 626)** The Spearman rank correlation coefficient

Enter the data into the MINITAB Data Window. Enter the number of Males into C1 and
the Females into C2. To rank the data values, click on **Data → Rank.** On the input
screen, you should **Rank data in** C1 and **Store ranks in** C3. When you click on **OK**, the
ranks of the Male enrollments should be in C3. Name C3 "Male Ranks".

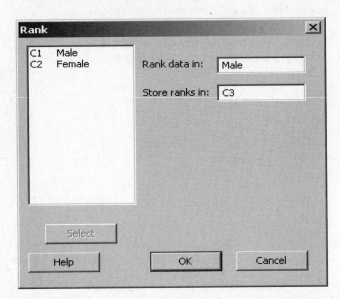

Repeat this using the Female enrollments and storing the ranks in C4. Now you should
have the ranks of the data in C3 and C4. Name C4 "Female Ranks".

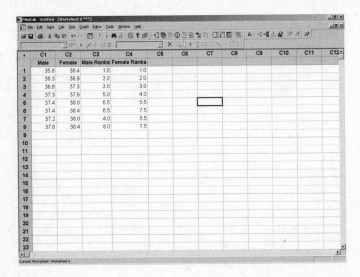

To calculate Spearman's rank correlation coefficient, simply use Pearson's correlation on the ranks of the data. Click on **Stat → Basic Statistics → Correlation.** Enter C3 and C4 for the **Variables** and select **Display p-values.**

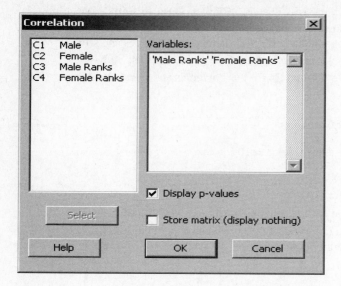

When you click on **OK**, the results will be displayed in the Session Window.

```
Correlations: Male Ranks, Female Ranks

Pearson correlation of Male Ranks and Female Ranks = 0.933
P-Value = 0.001
```

In this example, notice that the Spearman's Rank Correlation Coefficient is 0.933 and the *p*-value is .001. Since this P-value is smaller than α = .05, you would reject the null hypothesis. Thus, you can conclude that there is a significant correlation between male and female enrollment from 2000 to 2007.

◀

▶ Exercise 8 (pg. 629) Is vacuum cleaner quality related to price?

Open worksheet **Ex11_4-8** which is found in the **ch11** MINITAB folder. The overall
score is in C1 and the price is in C2. First, rank the data. Click on **Data → Rank.** On
the input screen, you should **Rank data in** C1 and **Store ranks in** C3. When you click on
OK, the ranks of the Overall Scores should be in C3. Repeat this for the Prices and **store
ranks in** C4. Now, calculate the correlation coefficient of the ranks. Click on **Stat →
Basic Statistics → Correlation.** Enter C3 and C4 for the **Variables** and select **Display
p-values.**

Correlations: Score Rank, Price Rank

Pearson correlation of Score Rank and Price Rank = -0.160
P-Value = 0.620

Notice that the Correlation Coefficient is -0.160 and the *p*-value is 0.62. Since the *p*-
value is larger than α=.10, you would fail to reject the null hypothesis. Thus, there is not
a significant correlation between Overall Score and Price.

◀

Section 11.5

▶ Example 3 (pg. 635) The runs test

Enter the data into C1 of the MINITAB Data Window. MINITAB requires numeric data for the Runs Test, so you will have to code the data. Click on **Data → Code → Text to Numeric.** On the input screen, you should **Code data from columns** C1 and **into** C2. Code **Original values** "M" as a **New** value of 1 and "F" as 2.

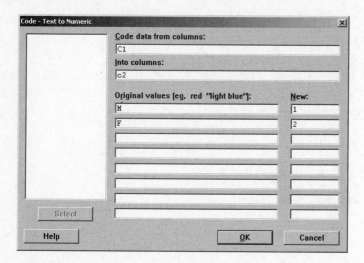

When you click on **OK**, the coded data should be in C2.

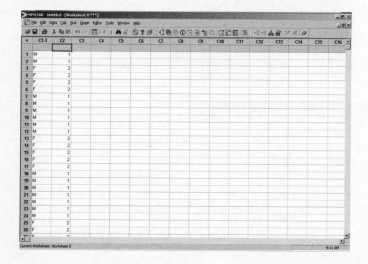

Next, click on **Stat → Nonparametrics → Runs Test.** Select C2 for the **Variable,** and select **Above and below mean.**

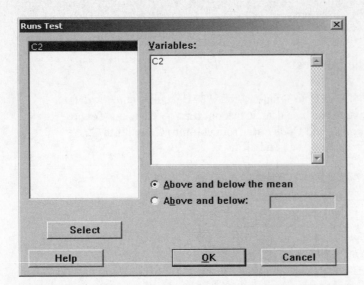

Click on **OK** and the results will be displayed in the Session Window.

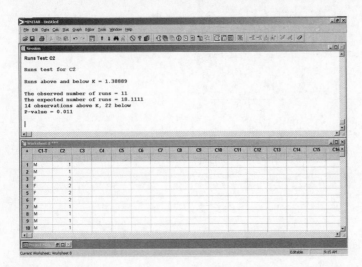

Notice that the observed number of runs is 11 and the *p*-value is 0.011. Since the *p*-value is less than .05, you can conclude that the selection of employees with respect to gender is not random.

▶ Technology Lab (pg. 647)   Annual income

Open worksheet **Tech11_a** which is found in the **ch11** MINITAB folder. The data should be in C1 - C4.

1. Construct a boxplot for all four regions. Click on **Graph → Boxplot → Multiple Y's Simple.** Select C1, C2, C3, and C4 for the **Graph variables.** Click on the **Labels** button and enter an appropriate title. Click on **OK** twice to view the boxplots.

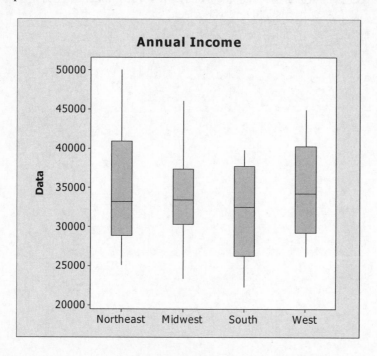

2. To perform a sign test on the data from the Midwest, click on **Stat → Nonparametrics → 1-Sample Sign.** Select South (C2) for the **Variable,** and enter 30000 for the **Test Median.** The **Alternative** should be **greater than** since the claim is that the median income in the Midwest is at least $30,000. Click on **OK** and the results will be displayed in the Session Window.

3. Recall that the Wilcoxon rank sum test is the same as the Mann-Whitney test in MINITAB. Click on **Stat → Nonparametrics → Mann-Whitney.** Enter Northeast (C1) for the **First sample** and South (C3) for the **Second sample.** The **Alternative** should be **not equal** since you are testing whether or not the median annual incomes are the same. Click on **OK** and the results will be displayed in the Session Window.

4. To perform a Kruskal-Wallis test in MINITAB, the data must be stacked in one column. To do this, click on **Data → Stack → Columns.** Choose all four columns to stack. Select **Column of current worksheet** and enter C5 and **Store subscripts in** C6. The subscripts will be numbers 1, 2, 3, or 4 to indicate which column the data

value came from. Be sure that **Use variable names in subscript column** is NOT selected. When you click on **OK**, C5 and C6 should be filled in. Click on **Stat → Nonparametrics → Kruskal-Wallis.** The **Response** variable is C5 and the **Factor** is C6. Click on **OK** and the results will be displayed in the Session Window.

5. Using the stacked data in C5 and C6, perform a one-way ANOVA. Click on **Stat → ANOVA → One Way.** The **Response** variable is C5 and the **Factor** is C6. Click on **OK** and the results will be displayed in the Session Window.

6. Open worksheet **Tech11_b** which is found in the **ch11** MINITAB folder. Repeat Exercises 1, 3, 4, and 5 using the data in this worksheet.

◀